Weeds and What They Tell Us

Books by Ehrenfried E. Pfeiffer

The Biodynamic Orchard Book

Pfeiffer's Introduction to Biodynamics

Weeds and What They Tell Us

You may also be interested in...

Biodynamic Beekeeping
Matthias Thun

The Biodynamic Farm
Karl-Ernst Osthaus

Biodynamic Gardening
Hilary Wright

A Biodynamic Manual
Pierre Masson

Companion Plants and How to Use Them
Helen Philbrick & Richard B. Gregg

The Foundations and Principles of Biodynamic Preparations
Manfred Klett

Growing Biodynamic Crops
Friedrich Sattler & Eckhard von Wistinghausen

Koepf's Practical Biodynamics
Herbert H. Koepf

The Maria Thun Biodynamic Calendar

Weeds

and What They Tell Us

Ehrenfried Pfeiffer

Floris
Books

Plant illustrations by Doris Clark, with the exception of
Mustard Family p. 15 and *Nightshade Family* p. 56 by Evelyn Speiden

First published by the Biodynamic Farming & Gardening
Association, Inc. as *Weeds and What They Tell*
This third edition published in 2012 by Floris Books, Edinburgh
in association with the Biodynamic Association. Sixth printing 2023

 Also available as an eBook

British Library CIP Data available
ISBN 978-086315-925-1
Printed in Great Britain by
Bell & Bain Ltd

Printed on sustainably
sourced FSC® certified
paper. Uses plant-based
inks which reduce
chemical emissions.

MIX
Paper | Supporting
responsible forestry
FSC® C007785

Table of Contents

Foreword

Ehrenfried E. Pfeiffer was trained as a scientist in Europe. Under the tutelage of Rudolf Steiner, he worked in the field of agriculture, at a time when European farmers were beginning to notice soil and crop deterioration, in spite of the wide use of mineral fertilisers based on the Liebig's Law (which said that growth was limited only by the scarcest resource). The farmers themselves were looking for a better way to restore their worn-out soils, and to refresh their faltering supplies of viable seed. It was Dr Steiner's feeling of urgency which intensified Dr Pfeiffer's efforts:

"The most important thing is to make the benefits of our agricultural preparations available to the largest possible areas over the entire earth, so that the earth may be healed, and the nutritive quality of its produce improved in every respect. That should be our first objective." (*Rudolf Steiner: Recollections by Some of His Pupils*, Golden Blade, 1958, p. 120).

When Dr Pfeiffer came to America, he established the Biochemical Research Laboratory in Spring Valley, New York. At the same time, he was putting biodynamic principles and preparations to work on his dairy farm in Chester, New York, which after seven years of biodynamic treatment, had reached the goals he had planned for it. Concurrent with managing the farm was his work on the sensitive crystallisation method of analysis for cancer and tuberculosis, and other extensive research for use by the medical profession. During the last five years of his life, Pfeiffer perfected a method of chromatography, which may be used to analyse the subtle differences in nutritional quality in various foodstuffs[1]. This method can be used to check quality, and to prove the value of different agricultural methods for food plants. It will be many years before scientists have thoroughly investigated the details of the methods which Dr Pfeiffer revealed, in essence and outline, both in Europe and in America. His intensive work in the laboratory was supplemented by farm surveys, field trips, biodynamic conferences, lectures on soils and agricultural methods, lectures on nutrition at Fairleigh Dickinson University and to the Natural Food Associates. Each year he led a small conference for farmers, held on an actual farm, for the purpose of helping biodynamic farmers with their problems. At these latter conferences, the highlight was often a trip to the various fields, where he took soil profiles and spoke extemporaneously out of his vast understanding of the geology, climate, and plants, as well as the insects and birds which inhabited that place. He had deep knowledge of the facts about each of these kingdoms – knowledge which he had augmented and substantiated many times by work in the laboratory. He took special care to point out the valuable qualities of certain plants, many of them weeds, as they grew in living relation to other plants.

This book presents one small segment of his knowledge of living plants: how they grow, what they reveal about their surroundings, and how their powers may be harnessed for the benefit of the human beings who appreciate and use them.

John Philbrick

Publishers' Note

Although Ehrenfried Pfeiffer's text has been edited into more modern English for this new edition, the reader should bear in mind that the book was originally written in the 1950s, and should be read and understood in this context.

Introduction

This book about weeds is by no means a complete description of all weeds, for there are more than four hundred here in America; *real* weeds, that is, which disturb our cultivation, farming and gardening. I have concentrated on the most characteristic ones of the north, the mid-east, and Midwest of America, and have tried to describe the properties which make them interesting to us. I have omitted their botanical descriptions, because they can be found in botany books, as well as in the very comprehensive *A Manual of Weeds* by Ada E. Georgia. For proper identification, the Latin names have been included. If you, the reader, don't feel comfortable with them, simply ignore them.

Methods of combating weeds are discussed in general (see both chapters 'The Battle Against Weeds'), and only in some instances are they repeated for each individual plant.

It is time for us to eliminate weeds from our cultivated lands. But we should also understand *why* we do it and *what* we're doing. Nature has a reason for allowing weeds to grow where we do not want them. If this reason becomes clear to us, we will have learned from nature how to deprive weeds of their 'weedy' character; that is, how to eradicate them from cultivated land, or rather, how to improve our methods of cultivation so that weeds are no longer a problem.

There is a significant problem, however: even if you try your best on *your* acre of land, there is often an abandoned place – a waste lot, a swamp or a wild area nearby – which spoils your land by windblown seed, by seeds carried by birds or other methods, undermining all your best efforts. Why not now, after the war in Europe has been won, spend a tiny fraction of what was spent on the war or post-war recovery, to start a national programme to combat weeds and insect pests? Only a large-scale operation will work, and it would be a valuable service to the country (as well as providing employment). I'm afraid, though, that politicians wouldn't like the idea, for there is no glory in it – only the gratitude of farmers and gardeners, and in any case they must learn to be content with whatever they get.

Weeds and What They Tell Us

Weeds are only weeds from our egotistical human point of view, because they grow where we do not want them. In nature, however, they play an important and interesting role. They resist conditions which cultivated plants cannot resist, such as drought, acidity of soil, lack of humus and mineral deficiencies, as well as a one-sidedness of minerals. They represent human beings' failure to master the soil, and they grow abundantly wherever people have made mistakes – they simply indicate our errors and nature's corrections. Weeds want to tell a story – they are nature's means of teaching us, and their story is interesting. If only we would listen to it, we could learn a great deal about the finer forces through which nature helps and heals and balances, and sometimes, also has fun with us.

Take, for instance, the common mould (*penicillium*). Nobody liked it, and when it grew on bread or cheese we were aware that things were getting old and not well-kept; but when penicillin was discovered, this cinderella mould became a highly worshipped princess. There is also the story of a gardener who had started a new garden on a ploughed-under alfalfa field. The following year, alfalfa was his nastiest weed, which he had to combat in order to grow peas, spinach and cabbage. The lush alfalfa almost outgrew the garden crops; nowadays, alfalfa is one of our most valuable farm crops, a fine soil improver. Therefore, we learn that 'weed' is a relative concept. A plant becomes a weed only through its position relative to cultivated areas. What we call weed may be a very lively, resistant plant, more vital than the cultivated ones, under certain growth conditions.

Consider also the case of sumac, with its many varieties which include poison sumac and poison ivy. It grows in abundance on swampy, wet and waste ground. Abandoned acres on hillsides, once cultivated, are gradually covered with it. It will infiltrate pastures, gardens and any place we would like to reclaim. In Europe, however, it is a decorative plant, favoured in gardens and parks for its exotic appearance. By no means would sumac in Europe dare grow as a 'weed' as it does here in America.

Weed groups

Weeds are specialists. Having learned something in the battle for survival, they will survive in circumstances where our cultivated plants, softened through centuries of protection and breeding, cannot stand up against nature's caprices. Weeds, therefore, may be grouped according to their peculiarities. There are three major and several minor groups. The major groups are our main teachers, indicating through their mere presence and multiplication what is wrong.

The first major group comprises weeds living on **acid soil** and indicating increasing acidity. To this group belong the *sorrels, docks, fingerleaf weeds, lady's thumb* and *horsetail* on slightly acid soil, along with *hawkweed* and *knapweed*.

The second major group indicates a **crust formation** and/ or **hard pan** in the soil. Here belong the *field mustard, the horse nettle, penny cress, morning glory, quack grass,* the *camomiles,* and *pineapple weed.*

The third major group consists of those weeds which follow human steps and *cultivation,* frequently spreading out with compost, manure and wherever people work the land. Here belong *lamb's quarters, plantain, chickweed, buttercup, dandelion, nettle, prostrate knotweed, prickly lettuce, field speedwell, rough*

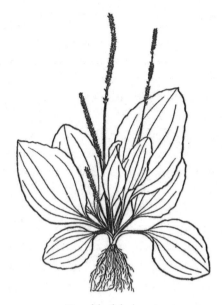

Broad-leafed plantain

pigweed, common horehound, celandine, mallow, carpetweed and other similar plants, all too frequent companions of our gardens and yards.

Minor groups consist of those which show up here and there (unless encouraged), and in fact are not always considered weeds. They are, more-or-less, an extension of nature into the realm of human beings. Here belong the many weeds of the *rose* family, indicating mainly a lack of care and cultivation. Here are also the 'pleasant-looking' weeds of the *pink* and the useful *legume* families, the weeds of the latter family preferring often light, sandy and poor soil, while the former thrive on rocky, gravelly soils, and along hedgerows and the edges of woods – a real 'borderline' group between cultivated and uncultivated nature.

The acid-soil-loving groups are our best warning lights, because they tell us exactly when changes begin in our soil. Acidity in soil increases with lack of air, standing water in the surface layer, cultivating of too wet a soil, insufficient cultivation, insufficient drainage, one-sided cultivation, and furthermore, the wrong kind of fertilisers, for instance, excess of acid fertilisers, increasing sheet erosion, and most important of all, loss of humus. Felt-like formations of roots and mosses will also appear on pastures.

We can frequently find acid-loving weeds even where there is natural limestone underneath, because the top layer of soil has been de-limed through one-sided cultivation and loss of humus, or excessive use of acid fertiliser. This happens when grain is repeated too frequently in a crop rotation; a long rest in pasture, alfalfa and clover is then required.

Typical weeds on **slightly acid soil**, due to insufficient cultivation, include *daisies, horsetail, field sorrel, prostrate knotweed*. For some of these cases, more frequent harrowing may solve the problem.

Very acid soil is mainly due to wrong cultivation and insufficient drainage. Typical weeds in this soil are *cinquefoil, swampy horsetail. hawkweed* and *knapweed* (also on 'wild' soils).

Salty soil weeds include *shepherd's purse, russian thistle, sea plantain, sea aster, Artemisia Maritima.*

A **hard pan** is formed when wet soil is turned by the plough, or standing water dries up in the surface layers. A hard crust is formed also through other errors in cultivation, mainly too wet disking and rolling, or when a soil dries up after having been cultivated before it has settled, or as a consequence of too deep ploughing. A **hard crust** also forms when fields are too frequently used to grow grain crops, with insufficient root and manure crop rotation in between. One-sided fertiliser application, particularly an excess of potassium, is accompanied by a group of weeds that includes the *wild mustard* and related weeds of the *cruciferae* family (though not shepherd's purse and Cochlearea officinalis) and the *horse nettle.*

Weeds of cultivation

Many of these weeds are real gourmands. They like soil which is hoed, manured or composted, and appear wherever land is being cultivated. In general they indicate a soil where the surface is loose, but where insufficient rooted organic matter is present. Here we find *chickweed* and *lamb's quarters*, *plantain* in trodden-down places, *thistles* spreading out in patches from wet pockets, and *stinging nettles*.

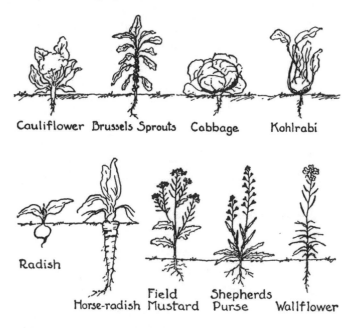

Cauliflower Brussels Sprouts Cabbage Kohlrabi

Radish

Horse-radish Field Mustard Shepherds Purse Wallflower

On **dry soils**, frequently with only a shallow layer of top soil and humus, or none: *mustard, silvery cinquefoil, Russian thistle, agrimony, catchfly, broom bush, crotalaria, dyer's greenwood, rabbit's foot clover, sweet clover, St Barnaby's* and *maltese thistle, common speedwell, prostrate pigweed,* most of the *spurge, shrubby St John's wort*

Peppergrass

Sandy soils: most *goldenrods, flowered aster, arrow leaved wild lettuce, yellow toad flax, ononis, partridge pea, broom bush* (the latter on submarginal land)

Steppe formation: *Russian thistle, sage, locoweed*

Alkaline soils: *sage brush, woody aster*

Limestone soils: *penny cress, field peppergrass, hare's ear mustard, wormseed, Canada blue grass, cornelian cherry, Barnaby's thistle, field madder, mountain bluet, yellow camomile (Anthemis Tinctoria)*

Excess potassium in the soil: *marsh mallow, wormwood, knapweed, fumatory, opium poppy.* Not a weed, but a good indicator, is *red clover*, which disappears with lack of potassium and increasing acidity.

Absence of lime: *yellow* or *hop clover, rabbit's foot clover, fox glove, wild pansy* (lawn!), *garden sorrel, sundews, white mullein, Scotch broom, black vetchling*

Gypsum soils: *common burdock* (neglected soils!)

Moist, badly-drained soils: *smartweed, mild water pepper, hedge bindweed, silverweed, white avens, swampy horsetail, meadow pink, hedge nettle, stinking willie, Canada* and *narrow-leaved goldenrod, tradescant's* and *purple-stem aster, joe-pye weed, marsh foxtail,* and *rice cut grass*

In grain fields: *wild buckwheat, all mustard, wild radish, penny cress, grass leaved stickwort, mouse-ear chickweed, morning glory, purple cockle* (these last two are serious warnings), *bachelor's button, tansy, poppies, chess* (also very bad).

On pastures: *buttercup, dock, knotweed, all fingerweeds* (the more the worse!), *white avens, grass-leaved stickwort, St John's wort, pokeweed, milkweed, wild garlic, briar, thistles.*

There are, of course, many more soil-indicating plants, but they are not weeds or intruders into our cultivated areas, but true wild plants.

Detecting soil types

To detect the properties of a certain soil, we have to look beyond the mere presence of one or other individual weed, and make judgements based on the prevalence of groups of weeds, as listed above. Although weeds of cultivation don't always indicate poor soil (although they should make us suspicious that soil might be dry or crusted, lacking in humus), an increased number of any weed – or of a distinct group of weeds – are always a warning of poor soil quality and cultivation. If 'wild' plants invade an area where they have not previously been present – bracken, for example – it's always a symptom of decline.

It is the increase of weeds of the acid or moist groups that is the most alarming sign, however, for an increase in soil acidity or neglected drainage.

We should also be aware of the *tendency* of certain weeds to appear and disappear, which can help us when judging a soil. For instance, the increase of weeds of the summer and fall (autumn) flowering class (see Chapter 'Summer and Fall-Flowering Weeds') is frequently a symptom of decrease of fertility and loss of humus (and lack of attention). If I had to select or judge a piece of farmland, I would spend time asking the local people about changes in the weed population. These biennial and perennial weeds show up also at winter time with their dried stalks or rosettes on the ground, detectable even through a thin layer of snow.

This booklet does not contain everything there is to know about weeds, or all weeds. Its purpose is to engage readers to notice and study the weeds around them, and find out about the most frequent or most important weeds. If readers learn to listen to nature's lessons – producing weeds in different soil, climate and cultivation conditions – then they have made the important first step in combating weeds: keeping them away from where they don't belong.

Remember: wherever weeds grow, they have something to tell us. Weeds are indicators of our failure.

The Battle Against Weeds:
Mechanical Warfare

Weeds are one of the most expensive crops to grow. They use up the light, moisture and nourishment which could be of benefit to cultivated plants. Removing weeds is expensive, no matter what method of destruction is used. It is estimated that weeds cost the American farmer and gardener more than $3 billion per year, which is as much as all other pests put together.

In small areas such as a front lawn or a small back garden, *manual combat* is possible: hand pulling is still the easiest and most effective method, supplemented by a pair of gloves, a hoe, a sickle and a small spade.

Other methods of eradication have to be adjusted to the peculiarities of the weed in question. It is not sensible to mow wild carrots on a pasture after the seeds have matured, nor to mow them when they just begin to blossom. In the latter case, they will form two to five new blossoms instead of one, and the mowing has to be repeated several times. The correct method is to let them grow, form blossoms, pollinate, start a seed formation and then mow. By this time the plant has exhausted the possibility of producing new blossoms and will die away.

The proper measure at the proper moment is therefore the key to controlling weeds to a bearable minimum – which should be considered a battle won. When Hercules fought the Hydra monster, he had a taste of what it meant to battle weeds – for each head cut off, three new ones grew. Only when he put fire to the sore and open neck did new heads not grow. However, we want to warn against the use of fire in combating weeds.

Wild Carrot

Many a barn, house, forest or valuable grazing land has been destroyed by this weed killer, and afterwards the weeds grew the better, encouraged by the ashes. Fire may also disturb the organic balance in the top layer of the soil, change its reaction, coagulate humus colloids, and produce the opposite result of what is desired. On a pasture with patches of weeds and briar, it has been observed that after the burning, the water remains longer on the burned-off patches than on the rest of the pasture. An increase in acidity resulted because of the lack of aeration, and many more and new weeds grew the following year.

The fight against weeds can be accomplished by means of small tools, mechanical cultivation of the soil, the kind of crops grown, and by means of chemicals. The main thing is to prevent roots, runners and rhizomes from spreading, and to avoid seed formation.

Eradication using small tools

The first step is to use a hoe to loosen the soil, which makes hand pulling of roots easier. The loose soil will encourage dormant seed to germinate, and the small plants can then be eliminated with a second hoeing. The most practical methods are:

1. surface hoeing
2. deeper hoeing and cutting
3. cutting and digging out of tap roots
4. raking off pulled weeds and little plants

Remember, the earlier the hoe is used, the less work there will be later on. In a garden, the spaces between rows and between plants should be kept clean from the very beginning. Once the weeds show up, it is often too late. Once weeds form a green cover, it is impossible to get rid of them at a reasonable cost. It's enough work trying to clear the many still or dormant seeds which are blown in. Pulling out by hand is impossible when the soil gets dry and hard because the taproot breaks off. Pulling should be done after rain, as long as the soil is moist.

There is a group of tall, summer weeds which can be mown with a sickle or scythe, or cut with a corn-cutting knife. All these weeds may grow until after blossoming, but should then be cut quickly.

Frequent cutting or mowing is necessary in order to starve the roots, taproots or rhizomes of biennials or perennials, that is, plants which would grow again from roots. Mowing is the only successful method on meadows or pasture where deep cultivation is impossible.

The following principles of mechanical warfare against weeds are the same for the garden as for the farm, although tools and machines used should be adjusted for scale.

Annual weeds

Annual weeds are those which grow from seed, and produce seed again the same year. These seeds germinate near the surface of the soil. A loose, crumbly soil will encourage them to grow which is what we want, as the small young plants are then easy to disturb by means of a cultivator or harrow. The best tool is the spring-tooth cultivator. However, before its application, a disc or spike-tooth harrow should be used in order to loosen the soil.

With spring moisture or rain, the seeds will germinate. As soon as they appear, use of the spring tooth will bring the entire plant to the surface, roots and leaves will wilt, and one set of weeds is thereby eliminated. The ideal conditions are first, to harrow or disc, then rain, then sunshine and dry weather when the spring tooth is applied. Should it rain after the weeds are uprooted, the roots may grow again and the whole thing starts over again. After the spring tooth harrow has been used, no other harrowing should be done unless the weeds are thoroughly wilted. The entire operation may have to be repeated if more seeds germinate. Two or three times tillage is therefore needed in case of many weeds.

Biennial or perennial weeds

Biennial or perennial weeds, which have rhizomes or tap roots in the soil from the previous year, are usually only damaged by a harrow. During a dry season this doesn't matter, but during a wet season this may encourage them. New shoots may grow from each cut-off part of the root or runner, so the well-intended measure helps to multiply rather than eliminate the nasty rhizomes. Therefore, a deeper working is necessary, first with the spring tooth harrow, or, if this fails, with a duckfoot cultivator (with a triangular blade) which digs up the root. Most garden tractors have excellent attachments, too.

Quack grass

A rototiller also takes care of the roots if used several times, going length and cross-ways. Roots are also cut by a lister cultivator. Most garden tractors come with very good equipment to use in between the rows. The lister equipment for farms is rather heavy: it needs a heavy tractor and does not work well in stony soil. In lighter soil with a hard crust formation, it might even be dangerous. It cuts the roots alright, but lifts the surface soil. With a hard crust and dry conditions, it may separate surface from bottom; the surface soil will dry out too quickly, and the cultivated plants will die. During the dry years of the dust storm period in the Midwest, much soil blew away for having been listered too frequently. All this mechanical equipment succeeds only if and when used under proper soil conditions, such as when the soil is not too wet and not too dry. A lot of experience is needed to do it right. This is why many farmers and gardeners are disappointed when – after much pain – the weeds still grow.

The best time for growing is late spring and early summer. For *oats, summer barley and wheat, grass and clover vetch, peas, spinach*, and small vegetables such as *radishes* and *spinach*, it can be hard to find the time to do all this preparatory work before sowing in spring. The fields and garden beds should, therefore, be made ready the previous year. Other crops favour extended cultivation before and after sowing. In the garden, such crops are beans, all kinds of *cabbage, cauliflower, peppers, tomatoes*; on the farm: *potatoes, corn, tobacco* – in general, any kind of bulky plant which is grown in rows, with sufficient space between rows for a machine. These crops are, therefore, our best standbys in mechanical warfare against weeds: they are planted later in the spring or summer, and in any case, need a significant amount of cultivation.

Many corn growers use checkrows, a system of cultivation which allows access to plants both longways, crossways and sometimes diagonally. It is starting to fall out of favour, but fields planted in checkrows are usually free of weeds, and ideal for any grain crop or grass and clover seed afterwards. In a garden, tomatoes, cucumber and squash could be planted in checkrows. Try it! Remember that weeds compete aggressively for moisture, nourishment and light.

Frequently, after crops have been harvested, an ugly-looking soil scar remains. This should be ploughed or disked immediately after the harvest, in order to get rid of the weeds. Eventually, these fields may have to be cultivated over again with a disc harrow, spike-tooth harrow, or with a duckfoot cultivator, until they're ready for fall (autumn) sowing or ploughing.

When all else fails

Now, what can we do if a field or a garden bed is so thickly covered in weeds that no tool will succeed? There is only one thing left: to plough, or in a small garden, to dig it over and cultivate for two to three months as often as you can manage. Don't be fooled that after ploughing and disking once, the situation is under control. In such a difficult case, you'll lose the crop anyhow, including the seed and labour on it, so why not have a thorough attack on the weeds first, and then enjoy much better growth next time, which will more than repay your effort. Such a system is called 'dry cultivation', or, 'follow with cultivation'. It increases the soil life, soil capillarity (free movement of water), and can be considered a soil improver.

After you are done with the cultivation, you can still put in a summer crop or a cover crop for late fall and winter, in order to prevent erosion. This is the same method as applied by our ancestors, when they let a field lie fallow during the summer, except that now the fallow land is cultivated. You will be surprised what a good crop will grow the following year. Try it on a small scale with the weediest patch on your farm or garden.

New gardens

A quick word about new gardens. You may want to change your garden space, for whatever reason. Perhaps there is an adjoining field or meadow, or even a waste plot suitable for your purpose. But, please, don't decide, after October, to plough up a place for your garden, as you won't have enough time to get rid of the weeds. Many of the seeds are dormant and are only awakened to full life by the ploughing. Thus, you will experience the full force of many years of dormant seed all at once, and will be forced to give up your garden before it is started. Plough it up in August or even earlier, and cultivate frequently in September and October. The soil then is much better fit for next year, and saves you a year's crop.

The war years, with the shortage of labour and high wages, have certainly been most helpful to weeds, and it will take some years to get the situation under control again. This is particularly true in the eastern states, where 1945 was an especially wet year.

The Battle Against Weeds:
Biological Warfare

You've already started biological warfare against weeds if you've tried to disturb their living conditions with cultivation, hoeing etc. when seeds are stimulated to germinate, and then uprooting. However, there are also three true biological methods:

1. choking out of weeds through faster and thicker-growing crops
2. elimination by planting other plants, which acts as repellants to certain weeds
3. changing the soil by means of manure, compost, lime and drainage

The most typical crops for choking out weeds are, in order: *soybean, sudan grass, buckwheat,* and similar fast growers. Buckwheat grows so thickly that it can become a weed itself if you let it go to seed. Still, since it grows on the poorest soils and collects lots of calcium, it will not only take away the light for low-growing weeds, but – if ploughed under as green manure – will sweeten the soil, thus fulfilling a dual purpose in rendering the soil more suitable for other crops.

Mustard can easily become a weed. *Shepherd's purse* is considered to be one already. However, both collect salt in larger quantities than any other plant. If grown on a salty marsh and ploughed under green, they will not only sweeten the soil, but take away the life element of the weeds ordinarily growing on such salty soil.

Along a creek, watery weeds, *swamp cabbage* and *docks* will frequently grow. The planting of willows and alders will improve the drainage, and gradually displace the distasteful swamp vegetation.

Rye is considered a good weed fighter, especially for low-growing weeds which survive the winter, such as *chickweed*. A grain crop, densely sown, will weaken *horse nettle*, particularly if followed with a hoed crop. *Wild carrots* will be replaced by *sweet clover*.

Quack grass can be choked out by sowing *soybeans, cow peas* and *millet*, if the land is first thoroughly cultivated, and the weather is hot and dry. Also, two successive crops of *rye* will choke out quack grass.

Papaver (poppy) and *wild larkspur* associate with winter, but dislike barley. *Charlock* and *field mustard* prefer oats. Their dormant seeds are awakened when oats are sown, though they do not harm oats. But they are very hard on rape and beets.

Plantain and *red clover*, as well as *dandelion* and *alfalfa*, frequently associate. I remember somebody in Kansas City complaining that suddenly dandelions had infested all lawns. At the same time, alfalfa was planted in the vicinity where it had never been grown before.

Camomile in small amounts stimulates the growth of grain. In large amounts, it is a disturbing weed.

Potatoes, with their intensive cultivation, usually oppress all weeds with large, extending roots. But *lamb's quarters* is particularly stimulated to grow with potatoes. So are some weeds of the buckwheat family. When a soil becomes infested with white or pitseed goosefoot, it's a good sign that a soil is tired of growing potatoes. Similarly, black nightshade often appears in soil which is tired of growing root crops.

Finally, we should also consider moss and lichen as weeds, which should be scraped off. If you want to avoid spraying

Bordeaux, on account of its copper sulfate content, you can make a semi-liquid paste of one-third cow dung, one-third clay, and one-third fine sand (you'll need to experiment a bit in order to find the right consistency), and then paint or plaster the tree all over with it. This will help the tree to avoid a recurrence, and will heal lesions at the same time[2].

Sprays and other treatments

A spray against poison ivy has been developed recently by Dupont and Company called *Ammonium Sulfamate*, which is harmless to other plants. It is sprayed at a rate of half a pound to a gallon of water for 100 square feet on the foliage of poison ivy (250 g to 4.5l for 10 square metres). The leaves wilt after some time (1-4 days), and do not come back if thoroughly sprayed. Some people think it might not be safe for the adjoining lawn, but we have not been able to obtain facts about probable harmfulness. Recommended[3].

Kerosene and crude oil: effective, but a sterile soil may result. Also, it is comparatively expensive. It fills the pores of the soil, covers the breathing openings in the skin of lower animals and affects soil bacteria and earthworms. However, California vegetable growers report the oil as successful against weeds in carrot and onion fields, but residues may give a kerosene taste to the carrots. In organic gardens, use with caution.

Common salt (sodium chloride): we don't like sodium chloride because of its damaging effect on most plants. It should be applied only where no growth is expected afterwards for quite some time. Cattle and sheep like to lick it. Concentrated brine can kill Canada thistle or quack grass, particularly after they are freshly cut, and when applied several times. It should be applied only in dry weather. Though cheap, the weeds do not think much of this weed killer.

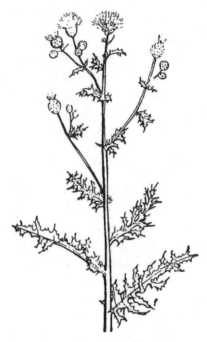

Canada thistle

Chlorate of sodium or potassium: very effective and kills everything. Therefore, it should only be used in places where plant growth is not wanted, such as paths, railroad dams, or similar. Some railroads have special spray wagons for it, but it may cause fires easily.

Sulphate of iron: used as a spray on foliage at the rate of 1 lb to 1 quart of water (500 g to 1 l) or 100 lbs to 50-60 gallons (45 kg to about 250 l) of water per acre – that is, rather concentrated. It is cheap and does not harm the soil, unless the soil is acid. In that case, sulphuric acid may form and cause damage. Grasses and grain seem to be resistant, according to A. E. Georgia, and it could be used against charlock, corn cockle, and other weeds in the grain field. It should be applied before the grain is heading and the weeds blossom. Georgia also says it is useful with peas but damages beans. Clover and alfalfa are blackened, but recover. It could be sprayed from a spray wagon like orchard sprays. Not to be used during a rainy spell. It will react upon common chickweed, buttercup and five-finger weed (cinquefoil). It seems to be particularly effective with wild mustard[3].

Copper sulfate: It is effective if used in a concentration of 1 lb to 5 gal of water (500 g to 28 l). The Cornell Station advises a 3% solution or 10 lbs of copper sulfate to 100 gallons of water (5 kg to 450 l), to be sprayed at a rate of 40-50 gallons (about 200 l) per acre, in a fine mist. It kills wild mustard, shepherd's purse (a weed which I would never specifically target, for it is very rich in minerals which may do a lot of good in the compost heap), white or pitseed goosefoot, ragweed, sow thistle, bindweed and dock.

Caution: copper sulfate is very damaging to beans, turnips and rape, as well as to earthworms. Many experiments have been done on composting grape vine leaves which have been sprayed with copper sulfate. Even when heaps have been properly built up and earth interlayered, fermentation was very slow, or none-existent, for many months. Earth coming from vineyards with lots of copper in it can be almost sterile for more than a year, and just lies inert, in spite of observation of all the other rules for composting. Due to its damaging effect on soil life, we caution against its use. Quite a few plants are also very resistant to copper.

Arsenic: Avoid!

A final practical tip: no matter what kind of spray you use, remember that large drops easily roll off the leaves of plants. A very fine mist sticks to the foliage much better and is therefore more effective. Adjust the nozzle of your sprayer accordingly, before you start. It is also ineffective to spray just before, during, or immediately after rain, so watch for the rainclouds coming or going.

Weedy Weeds

A typical family: buckwheat

Buckwheat is an interesting family. (The family name, *Polygonaceae*, is Greek – *polys*: many, *gony*: knots or knees.) Buckwheat itself is the tame brother of a group which is characteristic of sandy, light soils, as well as of acid soils, and includes *dock, sorrel, knotweed* and *rhubarb*. If soils are acid with standing moisture, then their weedy brothers will also invade clay soils. Wherever we see the high-knotted but erect stems, the red or reddish achenes (fruits) and scaly lobes of the dock and sorrel varieties, we can be sure that the soil is sour, has wet patches and insufficient drainage, as well as a hard pan. The spreading of these groups can be used to measure the increase of soil acidity.

Dock and sorrel

Patience dock (English spinach) has been grown in England and in Europe in gardens as a vegetable, for its slightly acid but refreshing taste. It prefers richer soils and grows near farmyards, manure heaps and roadsides. The *narrow-leaved, yellow,* or *curled* dock thrives in pastures, farmyards and moist places. The sorrels are frequently found in small quantities on pastures and meadows. They do not spoil the hay, although it is believed that the *field sorrel* is poisonous to horses. The young plants have been used in the past for savoury soup and as a vegetable. They are rich in oxalates, particularly the *tall sorrel*, the roots and leaves of which have pharmaceutical use. 100 lbs (45 kg) of

Curled dock

fresh leaves contains 4 lbs (2 kg) oxalate salts. The root stocks of docks are imported at a rate of 100,000 lbs (45 tonnes) a year, at a cost of half a million dollars. This money could help to get rid of a very common weed.

All are spread by seed, but the tall and deep-going roots of the docks and the rootstocks of the sorrels survive. The means of combating them is therefore frequent cutting, in order to avoid seed formation and to starve the roots. Deep cultivation in order to cut the roots is less successful. In small areas, pulling by hand, and/or digging deep with a spade may be helpful. The narrow-leaved dock has a widespread root system, and is very damaging to grain crops because it competes for moisture. Fields with insufficient tillage (if spring-tooth harrows are forgotten), and neglected pastures, as well as roadsides, barnyards, and edges of farm buildings and walls, are most apt to spread the seeds. Commercial seed, especially alsike (Swedish) clover seed may contain sorrel seed. Where there are docks and sorrels on a farm, we would try to improve the soil by reducing the acidity, by breaking the hard pan through frequent and deep cultivation, and proper aeration of the soil with a subsoiler or a spring-tooth harrow. Once the field drains properly, half the battle is won. Frequent cultivation will do the trick. Some experts recommend using lime, which in itself is correct against acidity. But if drainage and aeration of the soil are not taken care of, the lime simply sinks in the soil and is rendered worthless after a couple years.

On a pasture, mowing after pasturing is essential. Where there is increase of acidity on a pasture, cattle will not graze, and sorrels and docks will grow to seed unless frequently cut.

Chemical sprays, particularly the new hormone sprays, may be used on yards, near houses, and along the border of gardens.

Most of the docks like deep and good soil. They have deep-growing taproots, some two to three feet long, and grow up to six feet tall. As the Sorrels indicate more-or-less acid soils, their size is almost an indication of the degree of acidity.

Knotweed

Another group of the buckwheat family is the *knotweed* or *polygonum*, found mostly on acid soils. The *prostrate knotweed* (aviculare) is very common all over the world, frequent on garden borders, and along paths. The more trampled on, the better it grows. It is very rich in silica. All knotweeds are characterised by 'knots' on the stem from which branches come out. Some have white-pinkish blossoms. Others are the erect and the bushy knotweed.

The *swamp smartweed* indicates standing water and prefers moist pockets, though it grows anywhere, spreading out from low fields. It is nasty because of its creeping roots, which are easily broken through cultivation, and thus spread out. A typical

Pennsylvania smartweed

inhabitant of wet meadows and damp places is the *mild water pepper* (named for its bitter taste). Good drainage will eliminate it, as well as frequent mowing in order to starve the perennial roots. The *common smartweed* or *water pepper* has a very sharp juice that raises blisters on the skin.

Lady's thumb (persicaria), also called spotted knotweed, is a very troublesome, widespread fellow, growing in almost all crops and on meadows. It indicates a slight acidity, caused by insufficient surface drainage of the soil and lack of air. Frequent and thorough cultivation is therefore the best preventative. Its seeds are one of the more frequent impurities in red clover and it spreads easily where red clover is sown. The spotted knotweed is easy to recognise by its reddish stem and the lance-shaped leaves, pointed at both ends with a dark brown to black spot in the centre. The flower spikes are pink to purple.

Finally, there is *black bindweed* or *wild buckwheat*, and the climbing false buckwheat or *hedge bindweed* (P. convolvulus and scandens). *Wild buckwheat* is a moisture competitor in grain fields, climbing up and pulling down the grain by its weight. The seeds may be collected with the grain harvest, and spread out with chaff and manure. They resemble buckwheat seeds. The best way of combating them is by early cultivation with a weeder. After the harvest cultivation, follow with a root crop. *Hedge bindweed* grows mainly along hedges, is bird planted or creeps along the ground. It prefers moist soil.

In bygone times, all the smartweeds or knotweeds had value as secondary feeding plants, particularly in the Asiatic East. Some were used for tanning and dyes (brown, yellow and indigo). The Chinese smartweed was grown for indigo production. The Japanese developed, for feeding purposes, a giant knotweed with creeping roots, which in the second year already grew 6-9 feet (2 m) high. Folklore medicine used the twisted roots of meadow knotweed (bistorta) as a remedy against snake poison and against goiter (a thyroid problem) in horses. The common smartweed was used to raise blisters on horses. All this would indicate that this family contains quite a vitality which for some reason has never been properly developed and made useful, except in the case of buckwheat. Therefore, it became a weed.

A prolific family: cruciferae or mustard

Some of the most useful, as well as some of the most troublesome, plants belong to this family. Cabbage, mustard and shepherd's purse are all members. The mustards include *charlock* or wild mustard (Brassica arvensis), *Indian mustard* (B. juncea), *ball mustard* (Neslia paniculata), *wild radish* (Raphanus raphanistrum), the *peppergrasses*, common (Lepidium virginicum), green-flowered (L. apetalum), and field peppergrass (L. campestre), all of which are prolific inhabitants of our grain fields. Only frequent cultivation, harrowing before sowing and after harvesting, and weeding with a weeder and rotary hoe can keep up with them. The seeds of some, like those of the charlock and wild radish, can lie inert in the ground for 50-60 years. It happened that an old pasture was ploughed up. Nobody could remember if it had ever been used for grain, and the next year everything was yellow from mustard.

The more often grain follows grain in a crop rotation, the more weeds there are. Some of the mustards, *false flax* (Camelina sativa), *ballmustard, sand rocket,* the *hare's ear mustard,* the *tumbling mustard* (Sisymbrium altissimum), *wormseed mustard,* and *green tansy mustard,* most of which were imported with European seed, have spread out widely, particularly in the wheat-growing states of the Midwest and Canada. By using proper crop rotation, with no more than two grain crops every

Wild radish

five years, they may be gradually suppressed. In general, this group of weeds indicates a hard pan, as well as a surface crust formation in fields. They may spread out to and from the roadsides and waste ground, and should therefore, at their first slightest appearance, be hand-pulled. They pull easiest after rain when the soil is moist.

The wild radish in particular spreads out when soils are run down with too many grain crops and insufficient legume crops. Particularly where manure is scarce and potassium fertiliser plenty, it will grow abundantly, especially in wet years. However, cattle like it and it produces a good honey, as well as oil from the seed.

Penny cress (Thlaspi arvense) is likewise prevalent in the grain states. Like the shepherd's purse, its seeds contain 20% oil which, if ground with the grain, spoils the flour. There also exists a curiosity, the *mountain penny cress* (Thlaspi alpestre var. calaminare), that prefers soils containing zinc.

Field peppergrass, penny grass, hare's ear mustard, and bitter wormseed prefer soil with a high lime content, while the *bulbous cress*, a native of wet meadows and along streams and ditches, prefers an acid siliceous soil.

Winter cress (Barbarea vulgaris) and *scurvy grass* (B. verna) have both been cultivated as winter salad. They grow even underneath the snow, and can be considered as escaped weeds. They prefer moist places and are very much liked by cattle. Winter cress is also known as Barbara's cress because it still grows on St Barbara's Day, December 4th.

Shepherd's purse, which grows almost everywhere, finishes this group. As mentioned elsewhere, it is very rich in minerals and should, like its brothers and cousins, be picked early enough to become a valuable enrichment of compost heaps.

The Cruciferae family offers an interesting study in plant metamorphosis, if we put one species alongside the other, beginning with cauliflower, broccoli, cabbage, kohlrabi, radish, mustard, and ending with the toughest weeds. At one end we find the familiar round, head-shape, and the fine shape of the medicinal species; in the middle, the sharp-tasting mustards; and the rest at the other end. We can almost guess, without knowing too much about them, which plant belongs where, just from the shape and taste.

Morning Glory & Co:
the Convolvulus Family

This family are the real bad boys in our fields, gardens, hedgerows and fences, and in cultivated areas. Worst are the creeping and climbing species: *field bindweed*, or *morning glory* (Convolvulus arvensis), with long creeping, brittle roots, spreading over plants wherever they can. This, and the *hedge bindweed* (C. sepium), are a real pest in grain fields. Usually, too wet ploughing or disking of a field causes sticky soil conditions and a crusty formation which aids these weeds. Much cultivation, particularly with the duckfoot type of weeder and spring-tooth harrow will help, as well as thick-growing crops like soybean, vetch, and field peas – ie. crops which, by means of their roots, maintain humus and produce a crumbly soil. In persistent cases, temporary pasture may solve the problem, instead of aerable tillage.

And then there are the parasites – feeding and climbing on their hosts – the most dangerous of all, the *clover dodder* (Cuscuta epithymum), introduced with bad clover seed. Whenever it shows up in a clover field, it is best to give up growing clover and alfalfa for at least seven years (as long as the seed remains alive in the ground). When they first show up in small patches, it is worth cutting the infested plants and burning them after they have dried, on the very same place. The parasite eventually kills the host, but from the tiniest bit of stem another dodder can grow. It is recommended to plough under the crop before the seed ripens. Being a parasite, the full-grown clover dodder has no leaves, but is easily recognisable by its reddish stems and the cluster of white and sometimes slightly pink blossoms.

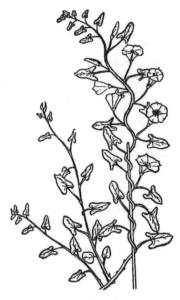

Bindweed

Flax dodder grows on flax and alfalfa fields. Infested fields should never be used for seed production. The best protection against dodders is clean seed from a reliable seed house, and a substantial fine for the one who sells impure seed. The seeds are very small – $\frac{1}{20}$ of an inch (1 mm) in diameter or smaller – smaller than clover seed. In the soil, roots germinate and develop as in a decent plant. But once the little seedlings have reached a host – roots – leaves are abandoned and the sucking starts. The *field* and *common dodder* (stem yellow, resp. yellow-orange), do not care who they suck, and are not so aristocratic as the clover dodder, which focuses on the most nourishing host.

To the convolvulus family belong, surprisingly, some beautiful garden flowers: the sweet potato (Ipomoea batata) and the rosewood, a native shrub of the Canary Islands, used for perfume and fine cabinet work.

Wild sweet potato vine (Ipomea pandurata) has a large, edible root deep under the ground, and grows as far north as eastern Canada, contrary to the *southern sweet potato*. If you want to surprise your children, have them dig out a wild sweet potato vine, or a *bryonia alba* (of the bluebell family, sometimes found in medicinal gardens, but not quite a weed yet in this country), where there appears on a tender, climbing vine quite a body, which is ten times and more the weight of the green parts.

The Goosefoot or Chenopodiaceae Family

If you were to send a child out into the yard, the waste lot, or to the roadside, and ask them to bring back a weed, they would almost certainly bring one or other of the goosefoot family, *lamb's quarters* (*white* or *pitseed goosefoot*) or *smooth pigweed* (Chenopodium album), or *oak-leaved goosefoot* (C. glaucum), or *spreading orache* (Atriplex patula). These are among the most enduring annual weeds and produce a tremendous amount of seeds, able to survive dormant for many years. Wherever human beings have cultivated the earth, these plants have shown up. They can stand quite a bit of drought, thrive on the poorest and most crusted soils and, even should we succeed in pulling them out completely in our garden, they will blow in again from the roadside, the waste plot, and manure and compost heaps. Spreading out compost means spreading out pigweed and goose-foot. They are detrimental to the compost because, with their roots, they absorb all moisture from inside the heap and bring fermentation to a halt.

It is essential to use clean seed, because the seeds of grain, clover, and alfalfa often contain impurities.

In addition, there is the *jerusalem oak* or *turnpike geranium* (C. botrys). It contains an aromatic oil, formerly used for rheuma-tism. *Wormwood* (C. ambrosioides), also called Mexican tea or Jesuit tea, was originally cultivated because of its medicinal value (teas for chest colds), but escaped. In East India, white goosefoot was cultivated as a vegetable. *Chenopodium quinoa*, though not grown in this country, has been grown in the high Andes, at a height of 12,000 ft (3500 m), as an important substitute for rye and barley, which do not grow at that altitude. The seeds are boiled in milk or roasted and used as cereal, and the leaves are cooked like spinach. The whole plant ranks high in value in the same class as potatoes, corn, and wheat.

We often feel that goosefoot was destined to play a better role in human life than to become an obnoxious weed. They appear wherever humans are, showing their inclination to be domesticated. Future plant breeders may possibly develop new cultivated varieties out of this family, long after our present cultivated plants have degenerated, for it is their extreme vitality and perseverance to grow that makes the goosefoot family so interesting.

Russian thistle

The *salsola kali* or *Russian thistle*, native to Russia, is not a thistle but another goosefoot, and is a typical steppe plant growing on salty borders. It was planted in order to collect soda near seashores on the Baltic Sea, as well as on the Adriatic Coast near Trieste. Imported into America with flax seed, it spread out as a troublesome weed in the western fields and prairies. Farmers consider it a crossbreed between thistle and cactus, which speaks for itself. This weed absorbs much potassium out of the soil and would be excellent on compost heaps. It could be used to de-salt the soil, but should not be allowed to seed. It also collects sodium salts as well as potassium (it has this property in common with certain orache (atriplex) species), when grown on salty marshes, near seashore, or on inland water places where evaporation is faster than drainage. This capacity means it thrives in soil that has lost its humus balance, and is beginning to transform itself into steppe-like conditions, containing great amounts of free-soluble salts, which may be otherwise washed out and become completely lost. Its appearance, therefore, should be a serious warning to the farmer, and a challenge to check on the condition of the soil.

The Parsleys: a Manifold Family

The *Umbelliferae*, or parsley family, is one of the largest families, with over 1500 species. All contain some kind of aromatic oil; many medicinal herbs, spices, vegetables and some poisonous plants are to be found here. The blossoms of this group are rather uniform so they may be mistaken one for the other. The *poison hemlock* (Conium maculatum) may be dangerous, for it contains a powerful poison, as experienced by Socrates. However the poison hemlock, and even the poisonous *water hemlock* (Cicuta maculata), a native, usually grows in swampy places and ditches, not on cultivated ground. In a pasture, it would be an indication of very bad drainage.

The most common weed in this family is the *wild carrot* (Daucus carota), not necessarily indicating bad soil, because its deep taproot normally indicates a deep soil, worthy of better treatment and better crops. Preventing this plant from seeding is best. The plant should be cut shortly after pollination and very close to the ground. If cut too early, many plants instead of one will spread out from the root. The size of the wild carrot, like that of many other parsley members, is almost an indication of the soil fertility. A rich crop of this weed is therefore a rather hopeful indication for soil improvement.

Famous edible plants and herbs in this family are celery (a salt plant), parsley, fennel, caraway, carrots, dill and chervil.

Fool's or dog's parsley is very poisonous, producing a burning sensation, nausea and mental disorders. We mention this plant because of its similarity to parsley. The antidote is vinegar or sour milk.

A very nourishing food plant with high yield even on poor soil is wild parsnip (Pastinaca sativa). It is its persistence to grow that makes it a weed, once you don't want it any more.

Cow parsnip (Heracleum lanatum) is one of the largest of the parsley family. Hercules, the Greek hero, is supposed to have discovered its medicinal value. Slightly poisonous, the root juice draws blisters. It appears very early in spring, when dangerous to cattle; it prefers moist soil and is found near springs. The unpleasant odour is very recognisable.

Meadow Parsnip (Thaspium aureum) has golden-yellow flowers, while most parsley members have white flowers. It occurs in moist pastures, and gives an unpleasant odour to milk in spring. Do not pasture wet meadows in spring. Its Mediterranean relative was famous as a remedy against approximately sixty different ailments.

Fortunately for our weed study, most of the weedy types of this family have not yet widely immigrated to America. Parsley weeds will develop, particularly on pastures and hayfields, where fresh manure and liquid manure are applied. They seem to do well on a lot of potassium and nitrogen. The application of well-rotted, composted humus on manure will avoid this eventuality. Cattle feeding on meadows infested with parsley weeds seem more inclined to mastitis, and even non-contagious abortion, than usual (I observed this with Swiss cattle). The more domesticated members of the family will do better the more a soil is cultivated, as every gardener knows.

Cattle usually avoid poisonous plants when grazing, warned by their natural instinct and noses. However we have seen that cattle, kept a lot in barns and fed on concentrates, lose this instinct and are apt to be more easily poisoned. Poisonous weeds in hay are dry, and have frequently lost their repellent smell. Cattle are then rather helpless, left with only prickly weeds in the manger. Old pastures and hayfields are particularly in need of severe control of damaging weeds.

The Plantains and Other Families (Lily, Spurge, Mallow, Grasses and More)

With their fleshy leaves close to the ground and long stems with just as long a spike carrying many seeds, the plantains are well known weeds on pastures and lawns, in gardens and on trodden paths, as well as along roadsides. They increase when standing moisture in the surface layers hardens the soil, but this indicates neither good nor bad soil, just that they like to be near it. Native Americans used to call introduced plantain the 'white man's footstep', including the *common*, or *broad-leaved plantain* (Plantago major), *narrow-leaved plantain* (P. lanceolata), and *hoary plantain* (P. media). There are also two natives, *red stem plantain* (P. rugelli), and *large bracted plantain* (P. aristata), the latter being perhaps the most pro-lific with one plant producing well over 3000 seeds.

All of them occur as impurities in grass and clover seeds. They contain a higher percentage of oil than almost any other seed. I was interested in this oil-producing quality (it is a very fine, almost tasteless oil), and I have cultivated the common and narrow-leaved plantains, with the result that spikes up to two feet long were obtained. Cattle will eat the plant, but we would not consider it for feeding. In England, however, it has been grown with clover in the last century and earlier, hence the infection of clover seed. Narrow-leaved plantain has been used with success as a home remedy, with compresses for bruises and strained joints. It also has a cooling and astringent effect when a few leaves are squeezed over a bee sting.

If it appears on a lawn, it should be dug out with a spade. If there are too many, it is time to plough and cultivate, in order to loosen the soil.

In Europe a species (P. maritima) grows near the seashore and near salt mines, absorbing salt and producing soda for which it has been used. It would be worth investigating whether it would be good enough for de-salting.

There are imported *poppies*, such as the *field poppy* (Papaver rhoeas), the long, *smooth-fruited poppy* (P. dubi-uim), and the *prickly poppy* (Argemone mexicana), and although they're not very widespread as yet in this country, they have the potential to become tremendous weeds. In Europe, the field and the smooth-fruited poppies pester all grain fields. Wheat infested with poppy weeds produces only small, light-weighted seed. Poppies are real robbers of the soil. They are grown for seed and oil, but the soil needs quite a bit of rest and reinforcement afterwards. However, this selfish property has an advantage – Dutch farmers grow a crop of poppies in order to choke out weeds which they could not get rid of by any other means. In gardens, poppies are grown for the beauty of their many-coloured blossoms. The seeds can lie dormant in the ground for years, and then show up again with a grain crop.

Great celandine (Chelidonium majus) is found here-and-there in barnyards, pastures and roadsides, and contains a yellow, slightly poisonous juice which was used against warts (hence the popular name wartweed, or killwart and devil's milk). It has also been used for sores of horses. Cultivation and early cutting before it goes to seed will keep it in check.

Purslane (Portulaca oleracea) is another widespread weed, found mainly in gardens and on good, cultivated soil. In England and Holland it is frequently cultivated, because it has a very refreshing, slightly-acid taste, and can be cooked like spinach.

Pokeweed

Carpetweed (Mollugo verticillata) is another low growing weed, forming mats in gardens and on paths, and thrives well on lighter, sandy soils. It will not resist hoeing and cultivation.

Pokeweed (Phytolacca decandra), a native perennial plant which starts along hedgerows and walls, and gradually infests pastures, is poisonous, but has medicinal value. It grows up to ten feet (3 m) tall, and is easily recognisable from the flowers in racemes, and the dark purple to black berries. Cut before it goes to seed. It loses its poison when boiled and has been used as vegetable in France and Northern Italy, where it was introduced about 1770. The red juice of the berries has been used in France and Portugal to colour wine and sweet treats.

Rough pigweed (Amaranthus retroflexus) and *spiny amaranth* (A. spinosus) is also called soldiers weed because it was supposed to have been brought North by soldiers returning from the Civil War. It is a native from tropical America, but is now seen almost everywhere. The seeds remain inert in the

Rough pigweed

ground for decades. Potatoes can almost be choked out by them.

The *prostrate pigweed* forms thick mats, and prefers light, dry soil. It is host to the white mould which damages beets, and should be eradicated by means of frequent cultivation and cutting whenever it shows up – on roadsides or waste places nearby. *Amaranthus* means 'not wilting', and it is indeed resistant to drought.

The lily family

Of the lily family we mention here only the *field garlic* and *meadowgarlic* (Allium vineale resp. canadense), the first introduced, the second native. Both are sometimes disliked, the imported one more than the native for its penetrating taste and odour. Ten minutes after cow or human has eaten field garlic, the entire body is already penetrated, and the breath smells 'garlicky'. After a half-an-hour, cow's milk is flavoured with it, and remains so for about 4-6 hours. Garlic cloves can be impurities in wheat which spoil the flour, their odour adhering to everything that comes into contact with them. The trouble is that with increasing years, the little bulbs grow deeper and deeper into the soil, becoming less accessible for cultivation tools and machines. Warfare should start early, by hand-pulling when the soil is moist, and by deep cultivation with a duckfooted machine or spring-tooth, in order to uproot the bulbs. If a pasture is very badly infested, nothing will help but repeated ploughing and a crop rotation with hoed crops until the evil is eliminated.

To avoid damage to milk, there is nothing left but to keep the cattle off the pasture, or to have them graze on garlic-infested pasture for a short time right after milking, and then to put the cows on another pasture towards noon, so that some of the odour can 'evaporate'. There is no doubt that garlic has a definite health value; medicines for high blood-pressue and scleroris are made from it. In old Egyptian documents, it was discovered that the labourers on the pyramid, more than 100,000 of them, were paid in garlic and onions.

The spurge family

Many of the spurge family have a preference for dry, light, sandy soils, but spread out in yards, gardens, pastures and other good land when given a chance. They have a sharp, milky juice with some poisoning effect. *Upright spotted spurge, hairy spurge, spotted spurge* (Euphorbia preslii, resp. hirsuta and maculata) and others are the best known and most widely spread. Many southern and western weeds belong to this family.

Leafy spurge (E. elsula) is an immigrant with creeping roots, which choke out everything else. So does a funny little plant called the *cypress spurge*, an escapee from gardens in the eastern states, with stems thickly covered with leaves (hence the name *cypress*). The milky juice of these two has been used against warts, and cypress spurge is employed in France as a laxative.

The spurge family is an interesting object for the botanist, with more than 3500 species, mostly tropical and subtropical.

Cypress spurge

The mallows – some of them escapees from gardens – such as musk mallow (Malva moschata) like cultivated soil and raw humus, and grow near houses and in yards. Some survive on dry, medium-to-heavy soils, like the *red false mallow* in pastures, and the *prickly sida* (Sida spinosa), originally a native from tropical America, now everywhere. *Bladder ketmia* or Flower-of-an-hour (Hibiscus trionum) is another escapee, now troublesome in gardens and grain fields. Some of the hibiscus group produce fibre: the *okra* of the south is edible, and the *cotton* plant is probably the best known of the mallows.

Common St John's wort (Hypericum perforatum) can be seen everywhere on pastures and fields except in the extreme north and south. It is a medicinal plant, containing a red oil which is used as a home remedy against bronchitis and chest colds. It has a strong, peculiar smell. The leaves have oily cells and look 'perforated' if held against the light. There is a *shrubby St John's wort* too (H. prolificum), found on dry, light, upland soil and submarginal land. The *dwarf St John's wort* appears on moist soils and waste ground. I don't mind these weeds, but if they grow out of control, they are certainly a symptom of neglected soil and insufficient fertilisation. The name *St John's wort* refers to the time of witchcraft: if collected during St John's Night (June 24th), it would protect against witches and evil spirits.

Grasses

Of the grasses a few are to be mentioned as more-or-less troublesome when not wanted, often in gardens and lawns. For all the creeping grasses such as *crab grass* (Digitaria sanguinalis) and *quack grass* (Agropyron repens), hand-pulling on lawns is needed before they form a mat, along with much and frequent cultivation in the gardens and much, much patience. Dry, hot weather will wilt the roots if they are brought to the surface.

Marsh foxtail (Alopecurus geniculatus) and *rice cut grass* (Leersia oryzoides) with knife-sharp leaf edges are invaders of our drainage ditches, and should be removed before seeding.

Chess or *cheat grass* (Bromus secalinus), a frequent impurity in rye and wheat, develops in wet years on grain fields, to an extent that farmers in bygone times thought their rye was transformed into cheat (which, of course, is not possible). Immediate cultivation after the grain harvest is necessary once the weed has settled; best is no grain, but cultivated crops instead, for one or two years.

Many other grasses of less value may show up in cultivated areas and pastures. Good basic farming practice, involving roper cultivation and sowing of clean seed, will help eliminate them.

Grazing sheep and goats after harvest on grain fields, abandoned fields, waste ground and secondary meadows can help eliminate the minor grasses.

Frequent intruders of cultivated areas are the *milkweeds* (Asclepias). They have medicinal value, and their floss or down became very important during the war. *Swamp* and *showy milkweed* grow on moist soils; the others prefer dryer soils. Cut them frequently to starve roots and hinder seed formation. Cattle avoid the acrid and bitter plants.

Hound's tongue (Cynoglossum officinale) typical of the *borage* family, has hairy leaves, a bad odour, nauseous taste, and burrs that stick to the fleece of sheep – which makes it hard work to get rid of in pastures.

The *mint family* is the true aromatic and herb family. If we were writing a book about herbs, it would get top billing. Any or all of this family can become weeds, but are relatively easy to check by means of cultivation. A few troublemakers can be mentioned, such as the *common horehound*, with hairy leaves and prickly calyx; *heal-all* (Prunella vulgaris), a tough fellow on lawns; and the prickly *hemp nettle* (Galeopsis tetrahit) and *hedge nettle* (Stachys palustris) on wet, badly drained meadows.

Poisonous Weeds

The nightshade or solanaceae family

The nightshades carry their poison in their physiognomy. They have, more-or-less, calix-shaped blossoms, and the deeper inverted the blossom, the more and stronger poison the nightshade has.

Potato flowers have no cup-shaped blossoms at all, and very little poison, as compared with the flowers of *tobacco* or *Jimson weed*. The visual thereby reflects the properties of the plant.

The most troublesome weed is the *horse nettle*, a native of America, also called *wild tomato* (Solanum carolinense). With its prickly leaves and its long, creeping roots, it is distasteful to gardeners and grazing animals alike. It prefers a crusted soil.

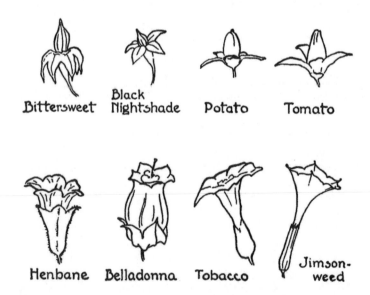

Bittersweet   Black Nightshade   Potato   Tomato

Henbane   Belladonna   Tobacco   Jimson-weed

Horse nettle

A crumbly structure produced by humus and frequent cultivation is its natural enemy. On pastures and meadows, it should be controlled by frequent cutting.

On pastures, yards and along fences, walls and hedgerows, the *Jamestown* or *Jimson weed* (Datura stramonium) will spread out, planted by birds. It is a very poisonous, bad-smelling plant, causing a kind of intoxicated state – but it has a certain medicinal value. In Arab countries, it is used to make a beverage from the *hairy thornapple* in order to enjoy lustful dreams. It was an ingredient of a medieval witch ointment, together with an extract of Atropa belladonna, a shrub. *Belladonna* is known for its poison, *Atropin*, used by eye doctors.

The third poisonous brother in witch ointments was hyoscyamus, the *black henbane* – an ugly looking, ill-scented plant with dirty yellow-blackish blossoms. Belladonna and hyoscyamus grow along roadsides, ditches and on waste ground. Children should be warned not to eat any of the nightshade berries, as some of their black or red berries look quite tempting. It is interesting to mention that in Asia and Europe, nightshades have spread out from Asia along the most travelled routes. These plants are really the travellers of the road sides and abandoned places.

Other nightshades are less troublesome, but are still bad weeds. The *black nightshade* has black seeds; the *bittersweet* tastes sweet at first, and then bitter. Both are typical weeds in gardens and in potato fields. They indicate a soil used up with excessive potato cropping and cultivation, a soil that is too loose.

The *petunia*, a native from South America with beautiful blossoms, is rather harmless. For a while it was a repellent of the potato bug. If it was planted between potatoes, such areas used to be avoided by the potato bug, but, alas, the bug got used to it.

Jimson weed grows frequently on or near compost heaps, testifying negligence on the part of the gardener. To combat it, pay more attention, pull it by hand, and never let it go to seed. 'By the sweat of your brow will you have food to eat' is literally true for the weeds of the nightshade family, as well as for their refined and cultivated brothers, *potatoes, tomatoes, paprika or spanish pepper; and tobacco.*

The true *mandrake*, with its alraun, or dwarf-shaped root, also belongs to this family of witchcraft – used by Circe – but does not grow in this country. What we call mandrake is a plant call podophyllum or *may apple*, growing on wet and swampy places.

The ranunculus or crowfoot family

An interesting example of the biological effects of one plant upon another is demonstrated by the *meadow buttercup* (Ranunculus acris). This well-known, yellow-blossoming companion of pastures, meadows and roadsides contains, like many plants of this family, a sharp, biting juice which is slightly poisonous, raising blisters on tender skins. Additionally, the buttercup contains a substance (probably ranunculin) which disturbs or even prohibits growth, particularly of clovers. On pastures we will therefore observe, that with the increase of buttercups, the clovers will disappear more and more. Early cutting will hinder the development of seed. Also, after grazing with cattle, sheep can be useful because they don't mind feeding on buttercups. Dried in hay, buttercups will lose their acrid properties. Once a buttercup infection has gone too far, ploughing and cultivating are the only means of combat.

The poison of a *mountain buttercup* (R. thora) has been used as arrow poison, and in days gone by it was used by beggars to rub it into the skin of children, in order to produce ugly and pitiful-looking sores.

Other varieties are the *early buttercup*, which prefers hillsides with increasingly shallow soil; the *creeping*, and the *bulbous buttercup*. The appearance of all of these should be taken as a warning signal that pastures and meadows need more care, particularly slight harrowing and compost.

Other more-or-less poisonous relatives are the *larkspurs* (Delphiniums), of which we like the garden larkspur, of course. There are the *field*, *dwarf*, and *mountain larkspur*, the latter frequently just called *cow poison* as the other name for the *purple larkspur* is just *poison weed*. They prefer high places or mountains, and can cause great damage to cattle and sheep. With increasing age, most of them lose their poison, so young plants are the most troublesome. Most of these species are more prevalent in the west and northwest of America. Their flowers are nice to look at, like their garden cousins, but beware!

Larkspur in fields can be choked out by growing barley (but it likes to grow in wheat).

The seeds of the more poisonous varieties have been used as an ointment against head lice.

According to mythology, larkspur grew out of the blood of a Spartan youth, Hyacinthus, who was accidentally killed by Apollo when he threw a disc.

The medicinal poison family: figworts and others

This family of figworts, or Scrophulariaceae, contains more-or-less medical weeds. That is, their 'poison' has a medicinal value, as in the *purple foxglove* (Digitalis purpurea), wanted in gardens and woods, but not in pastures. They like stony silica soils and raw humus.

The *mulleins* with woolly leaves and a long ever-growing stem covered with yellow blossoms are rather harmless, beloved remedies of the Celtic druids. (Old-fashioned midwives believed that a tea of mullein should be given to babies on arrival, 'To bring them down to earth' – this happened to me.) They grow on stony as well as good soil, are full of vitality, and indicate an absence of cultivation.

Yellow toad flax (Linaria vulgaris) has an acrid taste and long-running roots and grows on fields or meadows; it can be quite a disturbance. It prefers sandy soils and often grows along walls.

Common speedwell (Veronica officinalis) grows on dry, hilly soils, and is a harmless weed, formerly used as a medicinal herb, rich in bitter substances used to combat rheumatism and chest troubles. *Thyme-leaved ppeedwell* (V. serpyllifolia) is very common on wet, low pastures, a very modest plant indeed.

Purselane speedwell (V. peregrina) is used as a remedy against neck sores, hence the name *neckweed*; *corn speedwell* (V. arvensis) and *field speedwell* (V. agrestis) appear on cultivated ground and in gardens. All are liked by bees and are one of the earliest bee pastures in spring.

More poison: *American* hellebore (Veratrum viride) of the Lily Family – a native, not to be mistaken for the *black* and *stinking hellebore* of the Ranunculus family – is very dangerous to grazing animals. The rootstock, however, is used for medicinal purposes. Since it prefers wet places, its elimination by drainage and cultivation should not be too difficult.

The *poison sumac* and *poison ivy* of the Cashew family are so well-known that it is hardly necessary to mention them. They are invaders of the borderline of cultivated areas, and their poisonous acid can produce dreadful blisters on sensitive skin. The new hormone sprays and ammonium sulfamate may reduce them to unimportance – so we hope. Also, borax is used to destroy them.

Pleasant-looking Weeds:
the Rose Family

Most of the weeds of the Rose family look rather nice, with
handsome, bright-coloured blossoms, and as long as they stay
in their natural, wild origin – that is, on waste ground, roadsides, edges of woods, uninhabited areas and abandoned places
– they can hardly be called weeds. But when they migrate into
cultivated areas, they become troublesome. Moreover, they
prevail in pastures, meadows, or any uncultivated area, for they
are perennial. They are greatly disliked on pasture land, and
indicate that the land has not been cared for. The *sweet brier*
(Rosa rubiginosa or Eglantine or 'wild' rose) shoots up quickly.
It immigrates from hedgerows to pastures, which indicates
that the pasture has not been grazed sufficiently and should be
mown and harrowed. Due to their prickly canes, they can be
troublesome to cattle and sheep. Once they are established on
a pasture, the grazing animals avoid them and they then protect
the growing and seeding of other weeds until a real shrubbery
forms. Therefore, they are usually seen on abandoned farms.
Brier should repeatedly be cut on its first appearance, as long
as the canes are still soft, and the mowing machine or scythe
will take them.

Another companion of pastures and meadows are the *five-
finger weed* or *cinquefoil*, the *silverweed* (Potentilla anserina),
also *goose tansy*, the *rough cinquefoil* (P. monspeliensis) and the
silvery cinquefoil (P. argentea). All are natives and very persistent,
as long as the state of cultivation on a pasture is poor. They are
fairly alarming once they settle on a pasture. While the *silverweed* prefers moist and damp places, and therefore indicates

Rough cinquefoil

insufficient drainage, the others prefer dry and acid soils, and usually go hand-in-hand with the disintegration of humus and with increasing acidity.

Many cinquefoils, as the name indicates, have five-fingered leaves, which can be mistaken for strawberry leaves. On a pasture they are easily recognised, because they collect the morning dew in the centre of the leaf, glistening like a diamond. The leaves are close to the ground, thin runners spreading out in all directions. The flowers are bright yellow, shaped very much like a strawberry blossom, with five leaves arranged in the style of the rose family. The *silvery cinquefoil* has more of a greenish-yellow colour blossom with a white, woolly calyx. It will persist, even when all grasses are burned-up by drought. It has deep-growing, woody roots, which in the Middle Ages were used for tanning and red dye, as it is still done in northern Scandinavia. Once they show up on a meadow or pasture, the land should be harrowed on the surface in order to break up the felt of grasses and roots. The drainage should be checked, and the land should be thoroughly manured, or much lime-containing compost applied. Should the felty cover of weeds gradually choke out clovers and grasses, then there is nothing left but to plough, harrow and cultivate frequently, and start with a manure crop.

Only the *shrubby cinquefoil* (P. fruticosa) grows as a shrub several feet high if undisturbed, and chokes out everything else. In time, it forms very hardy, woody stems and rootstocks, which resist the scythe and mowing machine, and may have to be pulled or ploughed out with a team or tractor. Curiously enough, varieties of this cinquefoil are regarded as decorative garden plants, especially one variety from Nepal that has black-purple blossoms.

Wild strawberries can hardly be called a weed, even though they might increase on a pasture. However, they do indicate increasing acidity.

Another pleasant looking group of Rosaceae weeds is the Spiraea group, containing *meadowsweet* (Spiraea latifolia), *willow-leaved meadowsweet* (S. salicifolia) and *hardback* or *stee-plebush* (S. tomentosa). They are native perennials, propagated by seed, and have stems two to four feet (one metre) tall. Their flowers are in small, dense, terminal panicles, white or pink, and spire-like (hence the name spiraea), not at all resembling a rose flower, unless one analyses the little individual blossom. The leaves and flowers have a bitter taste, containing tannic acid, formerly used as a medicinal tea.

To combat: prevent seeding, close cutting, and pull out the roots (although, sometimes some look too nice to be eradicated).

An interesting weed that is found on poor, moist pastures and along streams is the *white avens* (Geum canadense or album). It is a native perennial and spreads out by seeds. The European relative of this weed was well-known as a spice, its taste resembling cloves. It has been used as a remedy too, and a distilled extract of the roots is used for fine liquours. The cross-section of the root shows an asterisk shape.

Here belong also the *agrimonies* (Agrimonia mollis, the soft; A. hirsuta, the tall hairy; and A. parviflora, the small-flowered). In bygone times, they were used as an antidote for various diseases and intestinal parasites, and also for yellow dye. Pliny tells us that King Mithridates (Eupator Dionysius) discovered the medicinal value of the agrimony.

The American varieties are natives, and propagate by seed; the *soft agrimony* also propogates by tubers. They grow along woodland borders and streams – the small-flowered variety favours sandy moist soil – while the soft agrimony prefers dry hillsides and woody borders. All three would be perfectly harmless were it not for the small, top-shaped burrs, with rows of hooks which stick persistently to the wool of sheep, much to the disgust of the wool trade. Unfortunately, they grow on poor, rocky and woody-hill pasture, which is good only for sheep.

To combat: early and frequent mowing, particularly along the borders from where they begin to spread.

Wild black cherry (Prunus serotina) and *choke cherry* (Prunus virginiana) can hardly be called weeds, for they grow at least into shrubs, if not large trees, the timber of which is highly regarded for cabinet making. Birds like the fruits, and drop the seed wherever they rest – on hedgerows, under electric wires – thus contributing to widespread propagation. The tent caterpillar puts its eggs on wild cherry, and produces the large, ugly tents we see so frequently along hedgerows. Many a farmer cuts down wild cherry hedgerows because of this. However, the eggs of the tent caterpillar moth can be detected easily. They are put in rings around the outer, smaller branches, and can be broken off in the wintertime. Personally, we like the wild cherry when grown to a tree. It is ideal as a windbreak in windy places, because it resists the bending effect of strong winds, as very few other trees do. The leaves of the young shoots coming out of the ground contain prussic acid, which may be harmful to grazing cattle. A. E. Georgia reports that half a pound of black wild-cherry leaves will kill a cow, so it may be better to burn the cuttings. However, cattle may have good instincts, and may not touch it. I have never encountered trouble with any poisoning effects on my cattle.

More Pleasant-looking Weeds: the Pink Family

To this interesting family of opposites, the Caryophyllaceae, belong some of the nicest garden flowers, such as the carnation, as well as some of the weediest weeds, such as the *cockle*, *sandwort*, and the *common chickweed*.

Common chickweed (Stellaria media) is the Number One Weed. It grows all over the world, and is so hardy that green leaves and even flowers may be seen underneath snow. (It is, in fact, not the only plant which grows underneath snow – rye and corn salad (lamb's lettuce) do, too.) Wherever soil is cultivated, particularly in gardens, on compost and in manure yards, it will appear. The only effective means of preventing its spreading is repeated pulling by hand. (A. E. Georgia recommends spraying iron sulfate on crops of strawberries, peas and grain)[3]. Birds are particularly fond of the buds and seeds. Pet canaries love the young plants. The weeded-out plants could be thrown into the poultry yard, or fed to the pigs, as they are very nourishing. Chickweed is a rather frail, weak-stemmed, branching plant with small leaves. The flower forms a white star. The seeds can remain dormant in the ground for several years. Chickweed has to be removed early, otherwise it will cover the soil like a felt blanket, hindering access of air and increasing the acidity of the soil. However, if removed before it goes to seed, it will itself produce a very good compost. See the end of this chapter for how to add chickweed and other weeds to a compost heap without spreading them with the compost.

While the common chickweed is an annual and creeps along garden beds, the *grass-leaved stickwort* is perennial and propagates by means of seed and rootstocks. It grows upright to about a foot (30 cm), on pastures, and along grainfields. It is a pleasant looking plant, like most of its relatives. Its white blossoms have five petals. Frequent cutting will starve the rootstocks.

The *fieldmouse-ear chickweed* (Cerqastium arvense) is a beautiful flowering, native plant used in gardens. It gets its name, mouse-ear, from the shape of its leaves. The blossom is large, white and star-shaped. Through its creeping rootstock (every joint can produce a new plant), it can become troublesome in pastures. Ploughing and cultivating may be the only solution. It prefers roadsides and sunny hills, and even grows way up on high mountains.

Common mouse-ear chickweed (Cerastium vulgatum), which propagates only by seed, has smaller leaves and blossoms than the fieldmouse-ear. It grows on fields and roadsides; it is a perennial and imported. Early cultivation in grain fields will lift up the shallow roots and eradicate it.

The *purple cockle* or *corn cockle* (Agrostemma githago) is a very nasty weed in the grainfield, particularly in winter grain. It is an annual and is propagated by seed. It mixes with the grain in harvesting and threshing, and spoils, due to its poisoning effect, the grain for feeding and flour. The seeds are black, somewhat similar to the caraway seed, and are particularly poisonous to sheep, pigs, rabbits, geese, ducks and poultry. The chaff from threshing should be checked for its presence before being fed to animals. The rather large flowers have five, reddish-purple petals, slightly mottled on the edge with dark spots inside. If grainfields become infested with purple cockle, hand-pulling may be necessary. Do not grow grain crops on such fields for several years, but plant crops which would be heavily cultivated – such as corn, potatoes, soybeans, vetch, or peas – can choke out the weeds.

The same rule applies to the plants which follow.

Campion (catchfly)

The *meadow pink* or *cuckoo flower* (Lychnis floscuculi) prefers moist meadows. It was introduced from Asia Minor and Siberia, and has value as an animal feed. Its blossoms are bright red (the well-known garden varieties are pink, white or blue, known as *phlox*). It has evidently derived its name, cuckoo flower, from the fact that it blossoms when the cuckoo calls, or, perhaps, because of the legend of 'cuckoo saliva' or 'spittle' from the Old World: the roots of all lychnis species contain saponin, which produces a soapy foam if stirred in water. Before the discovery of soap, together with the true saponaria, it was used for washing. To this same group (which does not grow in the southern states) also belong *red campion*, found on grainfields and pastures, and *white cockle* or *evening lychnis*, so-called for its white blossoms, which open in the evening and close with sunrise.

Another group, mainly infesting dry meadows, clover and alfalfa fields, which may become a pest if not controlled by early cutting include the *sleepy catchfly*, *forked* or *hairy catchfly*, and *night-flowering catchfly*. They are annuals or winter annuals, and propagate by seed. The *sleepy catchfly* has its flower closed most of the day, only opening in bright sunshine. The name, *catchfly*, is derived from a glue-like substance on the stem, which catches flies.

Weeds in the compost heap

Most people rarely put chickweed, thistles, goosefoot and others into a compost heap for fear of spreading it with the compost. We will, therefore, discuss a method to avoid spreading them out.

Green weeds and seeds will usually decay inside a compost heap as they are secluded from light, and to a certain extent, from air. Roots, especially rhizomes from which stems grow, will not always decay inside a heap. Seeds may germinate, runners may propagate, and rhizomes sprout on the surface of a heap, to a depth of several inches. This action is not bad, as on a compost heap it can be checked. If sprouting runners and rhizomas are disturbed at the proper moment, they may lose their energy.

In building up a compost heap, all the seedy weeds should be put into the interior part, way down near the bottom. Sprouting parts of plants and roots may go to the surface, roots on top, as they dislike light the most. As they grow, sprout, or germinate near the surface, a spade can be used to remove the upper layer, which is about 4-8 inches (10-20 cm) thick, and turn it back upside down. This procedure will usually check all seeds and most of the rhizomes and runners. Should this not be sufficient, turn the heap completely.

There can be many benefits of adding certain weeds to the compost heap, so persevere!

Summer and Fall-Flowering Weeds

The Compositae family: real weeds

Ten percent of all flowering plants, or more than 10,000 species, belong to the Compositae family (often referred to as the aster, daisy, or sunflower family). It's no wonder, therefore, that we find amongst them a huge variety, including medicinal and dynamic plants, weeds, and... more weeds.

The famous botanist Decandolle counted more than 600 plants of the sub-genus *senecio* (ragwort). We are mainly troubled by *common groundsel* (Senecio vulgaris) introduced from Europe, the native *butterweed* (S. glabelius) of the southern states, the *stinking willie* (S. jacobaea) growing further north, which is poisonous to cattle causing a hardening of the liver, and the *golden ragwort* (S. aureus) on moist fields. Though a few species prefer light soil, most of them have a liking for very well-cultivated soil. The battle against them is rather tedious, for their seeds are windblown.

A giant is the *great burdock* (Arctium lappa) with enormous roots, leaves and stocky burrs adhering to the hair of sheep, horses, dogs and to people's clothes. A. E. Georgia says the presence of this weed should be a disgrace to the owner – well, it is – but unfortunately, as with many of the biennial and perennial weeds, we just don't have the time. The *common burdock* (A. minus) is smaller but just the same. A large-seeded species, it grows on gypsum soil and may sooner or later show up on soils ruined by excessive application of sulfate of ammonia and lime (forming calcium sulfate in the ground). The burdocks are robbers of the soil. There is, however, a certain medicinal value attributed to their roots, for gout and skin diseases. 'Burdock Root Hair Oil', however, cannot possibly have an effect upon hair growth, for there is no oil or any distillable product in the root to make hair grow! Japan has produced a well-tasting edible burdock, apparently.

Thistles

And then there are the *thistles*. They are disliked because of their prickly leaves, though some have beautiful flowers – very handsome and proud looking. The biennial ones form a rosette of leaves on a big taproot and blossom only in the second year. All thistles are rich in potassium (use for compost!) and would have a high feeding value were it not for their prickles. In the nineteenth century, some people in Europe used to make a living by spudding thistles, and mashing and feeding them to pigs. Most of the Cirsium genus like deep and moist soil, though they may grow everywhere.

There is the *common* or *bull thistle* (Circium lanceolatum); *tall thistle*, mainly on roadsides and borders; *pasture* or *fragrant thistle*, very prickly; and the most damaging, the *Canada* or *perennial thistle* (C. arvense) with long, horizontal-creeping rootstocks. They take away food and moisture when in grain fields. Patches form in pastures which protect other weeds and therefore also spread them out. Much cutting, spudding, and deep-cultivation counteract this expensive crop. We have observed that when cut before the blossoms are open, many more will spread out from the rootstocks. It's better when they're cut after the blossoms are pollinated. But if, shortly after pollination, just the blossom heads are cut off, then the plant will bleed to death and wilt. But don't forget to do it!

A proud plant is the *Scotch thistle* (Onopordum acanthium), the heraldic plant of Scotland. It's edible and is sometimes cultivated for its medicinal value.

The *field sow thistle* (Sonchus arvensis) has deep-growing, creeping roots, and contains a yellow, milky juice, and grows well in wet fields. The *common sow thistle* (S. oleraceus) is edible, an excellent feed for geese and swine, and has been cultivated in Eastern and Southeastern Europe. The *spiny-leaved sow thistle* (S. asper) is the most prickly of the Sonchus genus. Hand-pulling when the thistles appear is still the best.

A valuable cousin is the *blessed thistle* (Cnicus benedictus) with medicinal and industrial uses, and a basic ingredient of Benedictine Liqueur and other bitter tonics (stomach bitters).

White or *ox-eye daisies* (Chrysanthemum leucanthemum), with long stems and well-known white and yellow centres, infest pastures, hayfields and lawns. Their seeds are frequent impurities in grass seed, but pass the digestive tract of animals unharmed, and are therefore easily spread out. However, they can be readily avoided, at least on a lawn (the daisy on an English lawn, growing close to the ground, is probably Bellis perennis) for they increase with increasing acidity of the soil, with standing surface moisture, with neutral humus becoming acid, and with the loss of lime. Good neutral compost with lime in it, bone meal, surface harrowing or frequent raking of the lawn, so that the easily forming upper crust and root felt are broken, will take care of the situation. Daisies are therefore a early warning sign: get ready and watch your soil. This is about the right moment to buy a soil-testing kit and assess the acidity.

The wild cousins of our edible lettuce are a real nuisance. First would be the *prickly lettuce* (Lactuca scariola) found on all cultivated grounds and abandoned fields; the *wild lettuce* or *wild opium*, with milky juice resembling opium (L. canadense); the *arrow-leaved wild lettuce* on dry, shallow soil; and a few other of the same genus.

A tall-grower is the *false dandelion* (Pyrrhopappus carolinianus) on dry soils, in Delaware and in the south.

The *hawkweeds* (Hieracium), of which the *orange hawk-weed* or *devil's paintbrush* (H. aurantiacum) and the more northern *field hawkweed* (H. pratense) are best known, spread out from wild places to fields and pastures. Found on acid soils, the hairy plant with its bitter juice is not touched by cattle. Most of the hawkweeds are invaders from Europe. The orange hawkweed is grown in gardens for its bright flowers. Folklore relates the name to the hawk, which is said to pick up the weed in order to strengthen its eyesight. Others believe the name is because the plant grows on rocky places where only the hawk can fly.

Beautiful plants with bright golden and yellow blossoms are the *goldenrods* (Solidago), with more than 60 native species. Some grow on, and are rather indicative of, dry soil, poor in humus. They can withstand long droughts and thrive on roadbanks and waste areas. Here belong the *gray goldenrod* (S. nemoralis) the *soft goldenrod* (S. mollis) the *stiff goldenrod* (S. rigida) *Canada goldenrod*, and the *narrow-leaved goldenrod* (S. graminifolia) with shallow creeping rootstocks, which can be found on rich and moist soil, in raw humus. Combat them by cutting them before they go to seed, and improve your soil with humus and better crops.

Asters

The *asters* are many. A. E. Georgia quotes ten, but there are about 160 native in North America. On moist and low-soil roadsides, we find *bushy aster* (Bohonia asteroides), *New England aster* with nice flowers, the *tradescanti aster*, the *willow-leaved aster* and the *purple-stemmed aster* (Aster puniceus) both on banks of streams and in swamps. If they invade your pastures and fields, it means drainage is needed.

There is also the *smooth aster* (A. laevis); the *white heath aster* (S. ericoides) with hard, bushy stems which spoil the hay, but are liked by bees; and the *many-flowered aster* which is typical of dry, stony, even shallow soils; some also grow on submarginal land, like the *Maryland golden aster*. Many of the asters have been developed into beautiful garden flowers. As weeds they shout at you, 'pay attention! Cultivate and improve your soil!'

There is a salt and soda collector growing on seasides and near salt mines, the *sea aster* (A. tripolium) and an alkaline indicator in the west of America, the very poisonous *woody aster* (Xylorhiza parryi) which grows in bushy clumps, smells bad, and tastes bitter as long as it is green.

Of the *fleabanes* (genus Erigeron) including the *Philadelphia fleabane* and *Canada fleabane*, the latter is a typical weed. It invades relatively good land. The Canada fleabane is one of the few 'presents' the American continent has offered to Europe. Around 1655, it was introduced to the old world in a stuffed bird, spreading out tremendously, particularly on stony soils. It just came in time for the last of the witch-craft ceremonies that were believed to cure chest illness and protect against the bewitching of children, and was probably used for incense. In this country it has a more rational use. The acrid oil is a repellant for mosquitoes, hence the name *fleabane*. Sensitive people may get sores, as from poison ivy, when touching the plant. It is still collected for medicinal purposes.

We mention the fact of this invasion of Europe because it often seems like a change of continent has changed the character of a plant, making it into a weed. In their native countries, the same plants have a tendency to 'behave'.

The *whitetops* also belong to the Erigeron genus, mainly invading meadows from waste ground. They mainly indicate a lack of attention and cultivation.

Camomiles

A little confusion exists about the *camomiles*. The real camomile is the *German* or *wild camomile* (Matricaria chamomilla) (see also the Chapter 'Dynamic Plants), recognisable by its fine, aromatic odour, and the hollow bottom of the blossom (cut the blossom with a sharp knife and you will see).

All the other camomiles, with solid blossoms, are not such dynamic plants, though they may be used as camomile for tea (such as *Roman camomile* (Anthemis nobilis)). Other camomiles are the *mayweed* (A. cotula) also called dog fennel, or *fetid camomile*, for its putrid smell, like dog's urine. Old-time beekeepers used to rub it into the skin to repel bees – even fleas will faint for a while. It is reported to repel mice, too, if rubbed into the floor and wall of the granary, and spread there.

The rest of the camomiles are of little importance, except that where they grow, better things will grow: the *field camomile* (A. arvensis), the *yellow camomile* (sometimes used as yellow dye), the *scentless camomile*, and the *pineapple weed* (Matricaria suaveolens), a western native, gradually invading the east of America. As soil-indicators on fields, they warn of crusted, hardening soil.

Of the *centaurea*, a rather widespread group, there is a 'thistle' type on roadsides, pastures and waste ground, the *purple star thistle* (Centaurea calcitrapa). *St. Barnaby's thistle* (C. solstitialis) seeds are frequently an impurity in alfalfa. *Maltese thistle* (C. melitensis) spreads out from seaports and mainly from California. Both immigrated with alfalfa seed to dry, hilly, and light soils.

A beautiful weed is *bachelor's button* (C. cyanus), also named *cornflower* or blue bonnets, and is found in grain fields. Contrary to other weeds in grain fields, it does not impair the growth of grain. Our experiments have shown that in small quantities, where it does not disturb as a moisture competitor, it even contains a growth-stimulating factor. On limestone soils the cornflowers are definitely blue. On acid soils, they frequently develop rose and pink flowers, and sometimes both colours on the same plant. We would say, the more inclined towards red, the more acidity is indicated for the soil. The *brown knapweed* and *black knapweed* (C. jacea resp. nigra) is found on acid soils. Medicinal values were once attributed to them, but have been forgotten.

The *chicory* (Cichorium intybus) has the colour-switching scheme from blue to rose, in common with the cornflower. It prefers a climate not too hot, and has a two-foot deep, fleshy, thick root which is used as coffee substitute. Several thousand tons were imported yearly which could easily have been found in one American county, like Orange County, NY. Chicory appreciates deep good soil so it is an indicator of good soil; the better it grows the better the soil is likely to be for root crops or potatoes. It grows along roadsides, dry ditches, fields and gardens. Its medicinal value, together with that of the dandelion root, was highly praised in cases of intestinal inflammation and for so-called 'spring cures'. Its close relative is the *endive* (Cichorium endivia).

Compositae plants with medicinal and other values

All the *wormwoods* have a very bitter taste and odour. Some have medicinal value against intestinal worms (containing santonin), increase stomach secretion, and aid in digestion. Also, *common mugwort* (Artemisia vulgaris), *biennial wormwood*, the most common, and the fast and tall-growing *annual wormwood* and *absinthe*, which grow in waste places, abandoned lots and stony, submarginal land.

Best known is *absinthe*, which goes into absinthe liquour and Vermouth. Their bitter taste and strong effect is well known, as well as their value in Hofman's Drops, a widespread stomach stimulant of our ancestors. A Vermouth tea-bath chases fleas from dogs and cats (we have tried it) and, it is said, keeps beetles and weevils away from the granary.

There is a maritime species, little known, which grows on salty soils and is worth investigation. The *common sage bush* (Artemisia tridentata) is one of the most drought-resistant plants on dry plains, western American deserts and ranges, and on alkaline soil.

Tansy or *parsley fern* (Tanacetum vulgare), so-called from the shape of its leaves, grows along fields, in dead furrows, and in yards, and has a very strong, bitter-aromatic odour. The Latin, *tanacetum*, derives from a Greek word indicating immortality, because the dry blossoms do not wilt. The distilled oil has been used as fly and mosquito repellant; medicinally the plant was used against intestinal worms (oxyuri) and in wine against stomach and intestinal spasms. The Russians used it as substitute for hops in beer, and rubbed the surface of raw meat in order to protect it against flies and mosquitoes. Also, rub your dog with it in August and see whether it chases the fleas away.

Tansy concentrates plenty of potassium and therefore belongs in the compost heap.

The *joe-pye weed* (Eupatorium purpureum) is not really a weed because it grows mainly in moist, damp thickets, ditches and streams, and only invades badly-drained meadows. It is named after an Indian herb doctor, but is also called *feverweed*, a name indicating its early use. It is said that its juice heals open sores and bruises, and hunters have observed that a wounded deer will search for it and eat it. The *thoroughwort* (E. perfoliatum) is closely related in properties.

Elecampane or *horseheal, horse elder* (Inula helenium), an introduced plant, is native to the eastern Mediterranean. The Latin name, *helenium*, derives from beautiful Helen, whose tears, according to mythology, were transformed into this plant when she was captured and brought to Troy. It has a thick, yellow taproot with a camphor odour, containing helenin and inulin, a form of carbohydrate, typical of some members of the compositae family. They are slowly-digested, and are used in dietetic preparations for diabetes, and also against catarrh of the lungs for their soothing effect upon the mucous membranes, and against the heaves of horses. Even a medicinal wine (Alant wine) was made from it, and Plinius recommended it to the Roman women to preserve their beauty – some weed! It grows in barnyards, along roadsides and in old pastures. The flowers are sometimes mistaken for *arnica*, bright yellow with dark center.

We almost forgot the naughty, useless *ragweed* (Ambrosia), a common pest of North America. Why it answers to the beautiful name *ambrosia*, the food of the Greek gods, only the gods know.

Ragweed

Good Weeds: the Legume Family

In the *legume* family, we do not necessarily expect weeds, for this group of plants belongs to among the most useful in nature, providing the soil with nitrogen, through nitrogen bacteria on their roots. Their root action and leaf decay also produce a very fine humus soil. The legumes, therefore, are our best friends and soil improvers. Many valuable plants, such as *alfalfa*, various varieties of *clover, beans, peas*, and *vetch* belong to it; also, shrubs – such as the *broom bush* – and the honey and black locust trees.

Let's start with *broom* (Sarothamnus vulgaris). It grows on the poorest stony, or sandy, slight to medium-acid soils. It is very rich in calcium carbonate, and therefore improves the soil through decomposition of its leaves and stem. Because it grows on the poorest, sandy soil, it provides excellent shelter for fresh tree-seedlings should one want to reforest a poor stretch, but when in excess, it may choke out the young seedlings. However, we do not believe it to be a weedy weed, just a weed.

Similarly, there is in the south the *crotalaria*, or rattle box. (The name is derived from the Greek word *krotalon*, meaning rattle; the loose seeds, when dry, make a rattling noise.) Its soil covering and improving qualities by far exceed its nuisance value, should it show up in an unwanted place. One variety (C. sagittalis) grows on bottom-lands in the Missouri and Mississippi basin, and is very poisonous to horses and cattle. It should be eliminated on pastures by cutting before seed or ploughing under, followed by cultivation of a crop such as cotton or corn, which requires repeated cultivation.

Ononis (Greek onos, meaning donkey, donkey food) favours and enriches sandy soil with potassium and calcium. On fern and heath land it is beneficial, but in cultivated areas it is a nasty weed.

Dyers greenwood (Genista tinctoria), also called dyers broom, was a very useful plant before the age of the chemical-dye industry. It was introduced to the New England home industry of weaving and dyeing (yellow and green) in the good old days. Bees are very fond of it, as are sheep and goats, although it is said that the bitter taste penetrates into the milk of cows. When the home industry disintegrated, it was not grown anymore and has now become a weed in the upper and drier hill-lands from eastern New York to Maine. Frequent cutting will gradually starve the roots to death. Dyer's greenwood propagates by seed and creeping root-stocks. This plant, like many others of the legume family, has been cultivated, and now is spreading out to waste ground or grows in places where no others can endure.

Clover

The same is particularly true of minor-value *clovers*. On a lawn, or in a flower or vegetable garden, they could not be considered 'damaging' weeds. But on pastures, meadows and in fields they frequently take space which otherwise could be used for more profitable higher yielding and more nourishing varieties. Once they grow out of control, frequent cutting is necessary in order to hinder their going to seed. Ploughing, cultivating, and resowing to better crops is the only other means of control.

To this group belongs *rabbit-foot clover* (Trifolium arvense). It grows on the stoniest, sandy and dry soils throughout the northeast of America. Its other names are *stone clover, pussy* or *hare-foot clover*, which indicate some of its properties. It causes intestinal disturbances to cattle and horses. There are also *yellow*

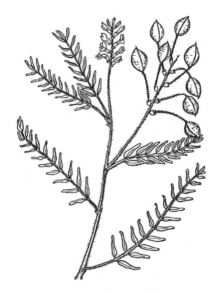

Woolly locoweed

or *hop clover* (T. agrarium) on roadsides and waste ground, as well as the *low hop clover* (T. procumbens) and the *sweet clovers* such as the yellow (Melilotus officinalis) and the white (M. alba). The latter was grown all around the Mediterranean Sea and is known there under the name of *Bokhara clover*. The seed was first imported from Asia and North America on sheep wool, to which it sticks easily. The seed can remain dormant in the soil for several decades. This plant grows under the poorest conditions and is therefore now very useful for reclaiming poor land. We used sweet clover as protection on slopes which were eroded, prior to starting reforestation. In this case, they will outgrow wild carrots and almost any other weed, giving the soil a firm hold, draining it and adding lots of humus. However, along highways and yards it may develop into a weed. Cattle graze it but do not like it. If used as hay, it has to be cured carefully, because it gets mouldy easily, and is then poisonous.

Another invader from Asia is the *birds-foot trefoil* (Lotus corniculatus). Its a modest fellow on waste ground, but when cattle get too much of it they bloat easily. The *burr clover* (Medicago hispida) is suitable for green manuring; its burr sticks to the wool of sheep, and spreads out over unwanted places. Through dormant seeds, it may show up as a 'surprise crop' after many years.

Wild sweet pea or *hoary pea* (Tephrosia virginiana), also called *Devils shoestring* and *catgut*, is a native that thrives on sandy soils and hill land. Its dense root-stocks are difficult to eradicate, and give it some of its names. It seems to contain a fish poison, but wild turkeys like it. There are a few native *wild beans*, too, the *trailing wild bean*, the *pink wild bean*, and the *small wild bean*, all having the Latin name *Strophostyles* (helvola, umbellata and pauciflora, respectively). All of these can be troublesome in cultivated fields. They are usually eradicated by hoe or cultivator.

One member of the legume family that is really troublesome is the *locoweed* (woolly or crazy weed), along with the *stemless* or *Colorado loco-vetch* (Astragalus mollisimus, resp. spicatus). It is a native perennial, and very poisonous to sheep and horses. It contains a nerve poison and drives animals insane. Although it does not grow in eastern America, it is interesting enough to be mentioned here. In the eighteen-eighties, the State of Colorado spent $200,000 on its eradication.

Partridge pea (Cassia chamoecrista) is another native legume which causes trouble to cattle and horses, although it is not near as bad as the locoweed. It causes scours whether green or dried in hay. This too, favours dry and sandy soils. It is more prevalent in the south and west of America, and grows close to the ground, spreading. It may infest pastures and meadows, and should be eliminated by repeated cutting before it goes to seed.

Dynamic Plants

The nettles

Though very widespread, the *nettles* (urticaceae) are rather unique, with only a few species belonging to the family. Distant cousins are *hemp* (Cannabis sativa) and *hop* (Humulus lupulus) and – believe it or not – the elm tree, the mulberries, the fig tree, and the sycamore. However, we are only concerned with the nettles. There is the *stinging nettle* (Urtica dioica) the smaller *burning nettle* (U. urens), the *tall* or *slender nettle* (U. gracilis) and the harmless, non-stinging *false nettle* (Boehmeria cylindrica).

The stinging agent is contained in fine hairs on the leaves and stems, and is mainly formic acid, and maybe a yet-unknown poison. Luckily, the worst stinging nettle of the Island of Timor does not grow here in America. Its burns last for many years, even a lifetime, the pain increasing with moist weather.

As weeds, we know these plants from gardens, compost heaps and waste ground, yards, roadsides – wherever dust and organic matter decompose, and the climate is not too hot and dry.

Nettles were known in oldest times for their medicinal value – they increase blood circulation and act as a stimulant. Very young plants are a well-tasting vegetable, prepared like spinach. They are excellent for chickens, and we have found that if shredded green, these young plants will be readily picked up by the chicks. Nettles are rich in nourishing substances and when used for feeding, they act as a preventative for diarrhea and deficiencies in poultry.

Now, what is the nature of dynamic plants, and why is the nettle one of them?

Dynamic plants are those which influence their surroundings in a specific way, so that other plants change their properties, or that a soil changes its character. But this influence has to go further than the mere competition between minerals, water and light. The stinging nettle has at least three properties which illustrate its dynamic character. It makes other plants grow more resistant; changes the chemical process in neighbouring crops; and stimulates humus formation. For example, experiments have been done growing stinging nettles in rows between tomatoes. It was observed that these tomatoes did not rot so easily, and kept their juice for four weeks, without preservatives, and without boiling. In a different experiment, peppermint was grown in three rows, then one row stinging nettle, then three more rows of peppermint, etc. The peppermint oil content was 2.5% in the plants near the stinging nettle, while the control plants without the nettle produced only ¾-1% of oil under otherwise equal conditions.

The third property can be studied if one digs out the soil close to the nettle roots and observes the kind of humus that is formed there – a blackish-brown, neutral humus. The leaves and stem of this plant rot to an ideal humus. There may be certain secretion on the roots which stimulates soil life and fermentation. The biodynamic method of gardening uses a humus made from stinging nettle (only the Urtica dioica will do) in order to stimulate fermentation in a compost or manure pile.

You might think that it's silly to cultivate or collect a weed. But don't forget that many weeds have medicinal qualities or poisonous effects. Producing neutral humus, to stimulate quality properties in other plants, can be considered 'medicine' for soil. All humus producers are good medicine for sick soil. Mustard, for example, has an alkaline secretion on its roots and can sweeten an acid soil. Mustard, shepherd's purse, and a few others absorb excessive salts and return them in organic form. It is a pity that this field of organic and dynamic processes has, as yet, been so little investigated.

Cultivated plants, frequently derived from weeds, have lost such dynamic properties in the process of learning how to grow in size, rather than in quality. This is true of cabbages, radishes and beets. A very surprising result can be seen if one transplants the beautiful Alpine *edelweiss* from its natural habitat into a garden: it changes into thick, dull leaves, more like a cauliflower tumour, not at all resembling its original beauty.

Returning to our stinging nettle: it is quite independent and at first resists 'proper cultivation'. You may have difficulty if you attempt to cultivate the stinging nettle in the same garden where it grows wild. We had great difficulties when we tried to grow stinging nettles for their dynamic properties in Egypt. By the way, the nettle has a rough but durable fibre, and textiles made of it keep the body nice and cool.

Other dynamic weeds

A most amusing plant for the student of dynamic properties is the *quack grass*, or *couch grass* (Agropyron repens). When Sisyphos rolled his stone uphill in endless repetition, as Greek mythology tells, he missed a far greater punishment, namely, weeding quack grass. Each broken-off piece of root or stem develops endless new plants. In botanical books, you read that it propagates by seed and by its creeping roots. The seeds, we have found, are frequently sterile or there are no seeds at all in the ears. Since this plant, curiously enough, is the nearest relative to wheat, we thought of transforming it into a useful plant. To our great surprise, this weed, full of vitality, resisted cultivation. If you force it to grow upright instead of creeping, it becomes a weakling and you have to apply tender care to make it produce fertile seeds. It took many years to gradually transform it into a straight, upright plant, producing about 500 seeds out of one plant, with no creepers at all. We transformed the weedy vitality into a useful property. In doing so, we also got deep knowledge and inside information about nature's laboratory. Couch grass is good cattle feed and would be good for covering gullies and road banks, particularly where live soil has been cut open – if only we could get rid of it after it has prepared the ground for other and better things.

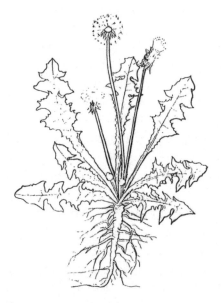

Dandelion

Another dynamic weed is the *dandelion* (Taraxacum offici-nale) which flowers first in spring and early summer – not the high-growing dandelion (Leontodon autumnalis), which flowers later and is sometimes mistaken for dandelion. This weed is somehow associated with clover and alfalfa, prefer-ring deep, good soils. Where it grows, the soil is good enough for the better legumes, and if they do not grow, it is not the fault of the soil or the dandelion, but yours. Earthworms like the soil around the dandelion, for this plant is another neutral humus producer.

There are people who go almost mad when they see dandelions on their lawn. If that's the case, they should take a spade and dig them out. But then they should also realise that there would not be a lawn if dandelions could not grow. Even if they grow in thick patches they are not in competition to grasses, because of their 3 feet (1 m) deep roots. The dandelion transports minerals, especially calcium, upward from the deeper layers, even from underneath the hardpan, which it penetrates, and deposits them nearer the surface. It 'heals' therefore, what the soil has lost, what has been washed downward. The dandelion grows too close to the ground to be harvested for compost, but it makes its own rich humus and when it dies, its root-channels act like an elevator shaft for earthworms. It's basically true that what the earthworm, as an animal, does to the soil, the dandelion also does as a plant. And when it really grows as a weed, it only admonishes you: 'Take care of your humus and your soil life'. If abundant on a pasture, it takes care of the vertical drainage, as long as you do not get active and do it yourself by harrowing.

The next dynamic principle is demonstrated by the *camomile*. This plant is a surface crust-breaker and its presence, our experiments have shown, stimulates wheat to grow heavier and with fuller ears. But it does this only when present in minute quantities, such as one camomile to one hundred or more wheat plants. (If grown in a large amount between wheat, eg. 20:100 or more, it will hinder the wheat development.) This is because camomile can cope with the hard crust that has developed on the soil, but wheat cannot. Camomile tells you, therefore, that you ploughed your field too wet and did not do enough tillage, and perhaps have used too one-sided an acid fertiliser. Manure and legumes in the crop rotation are then required. Excessive growth of camomile therefore warns you, in a friendly way, to change your rotation. If you do not change, you will encourage the wild mustard gangsters.

Camomile is very rich in potassium and calcium, and good for compost. In addition, its flowers make a valuable tea which has a soothing effect upon intestinal inflammations or sores, in man and animal.

We will end this chapter with the description of the equisetum or horsetail. But I want you to know that there are many more dynamic plants out there: go to the fields, and discover them.

The Equisetum or Horsetail: a Lone Ranger

This is quite an interesting plant, the final remainder of the huge trees of the carbon forests; propagating by spores and creeping rootstocks. The most common is the *field horsetail* (Equisetum arvense) which grows in sandy and gravelly soils, on a high ground-water level. Horsetails prefer a moist and cool climate, although they will survive a dry summer or a dry location, such as railroad dams. This plant is only troublesome when it penetrates pastures or fields; then it's an indicator of insufficient drainage and cultivation. Taking care of the soil also helps take care of the weed. Another variety is *swampy equisetum* (Equisetum palustre) which grows in very wet swampy places, in woods, and is worthless.

All equisetum plants, the field horsetail most pronouncedly, have a silica skeleton. Burning the green stems and leaves in a hot but quiet flame removes all organic parts, and leaves a white skeleton of silica which will still show the original structure of the little stems. It is this silica content (as high as 80% of the entire ash) which made the field horsetail a valuable tea remedy to our ancestors. Since the equisetum is very resistant to fungus growth, the biodynamic method uses a tea (a 0.5-2.0%

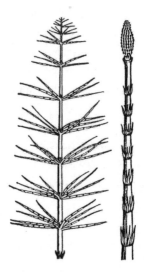

Horsetail

solution, made by boiling for 15-20 minutes) as a biological spray against mildew and other fungi, on grapevine, vegetable, rose and fruit trees. This tea is not as powerful as copper and arsenic sprays, but has a very soft and swift action which does not disturb the soil life. However, it should not be applied until the first cause of fungus infection – too moist a stand of the infested plant – is taken care of too.

Unique to Equisetum arvense is a distinct effect upon the kidney system. I have used it for kidney disturbances in cattle and horses, as well as dogs. Contrary to general belief, no poisoning effect was observed, even when as much as a pint of tea was used at one time. Because of its usability, horsetail has never been considered a bad weed on biodynamic farms, and it is collected wherever it can be found. Although the plant propagates easily once it is established, it is very difficult to start or propagate where it is wanted. It propagates by spores which are spread out from fertile brown, but leafless, stems, in springtime. These grow out of the rootstocks. The leafless, brown stalks are about 4-8 inches (10-20 cm) high and are frequently overlooked. All parts of the equisetum family, which grow above ground, have small joints connecting one section of stem or leaf with another. They pull apart easily, much to the delight of children (and adults) who often cannot stop playing with it until the whole plant is taken apart.

Before cleaning agents were manufactured, the boiled horse-tail plant was used to clean and shine surfaces of silver and pewter. It was, therefore, called *zinnkraut* in Germany.

The useful variety, arvense, differentiates itself very easily from the other useless varieties. These latter wear a kind of collar, formed by small, black-pointed leaves about the joints, while the collar of the true field horsetail is green or only slightly discoloured.

End Notes

1. See also: Marcel, *Sensitive Crystallization*

2. See also: Masson, *A Biodynamic Manual*, p92ff, 'Tree paste, sprays and pruning scars'

3. This product is not currently approved for use by organic growers. For up to date information on organic sprays, including those using sulphur, see Masson, *A Biodynamic Manual*, p145ff

Bibliography

Georgia, Ada E, *A Manual of Weeds*, Macmillan 1914

Masson, Pierre, *A Biodynamic Manual*, Floris Books 2011

Osthaus, Karl-Ernst, *A Biodynamic Farm*, Floris Books 2010

Pfeiffer, Ehrenfried E., *Biodynamic Farming and Gardening*, Portal Books 2021

Pfeiffer, Ehrenfried E. and Michael Maltas, *The Biodynamic Orchard Book*, Floris Books 2013

Pfeiffer, Ehrenfried E., *Pfeiffer's Introduction to Biodynamics*, Floris Books 2011

Index of Weeds by Common Names

Illustrations are shown in bold text